Information Circular 9509

Ergonomics Processes:
Implementation Guide and Tools for the Mining Industry

By Janet Torma-Krajewski, Ph.D., Lisa J. Steiner,
and Robin Burgess-Limerick, Ph.D.

DEPARTMENT OF HEALTH AND HUMAN SERVICES
Centers for Disease Control and Prevention
National Institute for Occupational Safety and Health
Pittsburgh Research Laboratory
Pittsburgh, PA

February 2009

This document is in the public domain and may be freely copied or reprinted.

Disclaimer

Mention of any company or product does not constitute endorsement by the National Institute for Occupational Safety and Health (NIOSH). In addition, citations to Web sites external to NIOSH do not constitute NIOSH endorsement of the sponsoring organizations or their programs or products. Furthermore, NIOSH is not responsible for the content of these Web sites. All Web addresses referenced in this document were accessible as of the publication date.

Ordering Information

To receive documents or other information about occupational safety and health topics, contact NIOSH at

> Telephone: **1–800–CDC–INFO** (1–800–232–4636)
> TTY: 1–888–232–6348
> e-mail: cdcinfo@cdc.gov

> or visit the NIOSH Web site at **www.cdc.gov/niosh**.

For a monthly update on news at NIOSH, subscribe to NIOSH *eNews* by visiting **www.cdc.gov/niosh/eNews**.

DHHS (NIOSH) Publication No. 2009–107

February 2009

SAFER • HEALTHIER • PEOPLE™

Contents

Acknowledgments

Abstract

I. Introduction: Ergonomics and Risk Management
        Basic Elements of Ergonomics Risk Management Processes
        Participatory Ergonomics
        Evolution of Risk Management Processes

II. Ergonomics Processes: Case Studies
        Bridger Coal Co.
        Badger Mining Corp.
        Vulcan Materials Co.
        Lessons Learned
        Summary

III. Process Effectiveness
        Bridger Coal Co.
        Badger Mining Corp.
        Vulcan Materials Co.

IV. Implementation Tools
        Tool A: Risk Factor Report Card
        Tool B: Musculoskeletal Discomfort Form
        Tool C: General Risk Factor Exposure Checklist
        Tool D: Ergonomics Observations
        Tool E: Handtool Checklist
        Tool F: Manual Tasks Risk Assessment
        Tool G: Ergonomics Task Improvement Form
        Tool H: Risk Factor Cards
        Tool I: Ergonomics Sticker for Mining Industry

V. Training
        Introduction
        Ergonomics and Mining: Ensuring a Safer Workplace – Training for Management
        Ergonomics and Risk Factor Awareness Training for Miners
        Ergonomics Observations: Training for Behavior-based Safety Observers

References

Appendix – Ergonomics Processes: Beyond Traditional Safety and Health Programs

Acronyms and Abbreviations Used in This Report

ACGIH	American Conference of Governmental Industrial Hygienists
BBS	behavior-based safety
CARE	Corrective Action Request for Evaluation
GAO	General Accounting Office
MSD	musculoskeletal disorder
MSHA	Mine Safety and Health Administration
NDL	no days lost
NFDL	nonfatal days lost
NIOSH	National Institute for Occupational Safety and Health
OSHA	Occupational Safety and Health Administration
PPE	personal protective equipment
S&H	safety and health
SHE	safety, health, and environmental

Acknowledgments

The authors thank the many organizations who helped to demonstrate that ergonomics can be integrated with existing safety and health programs to improve working conditions for their employees. Specifically, we thank Paul Gust, Kean Johnson, and Pat James of Bridger Coal Co.; Marty Lehman, Mellisa Stafford, Linda Artz, and Don Seamon of Badger Mining Corp.; and Dick Seago, Mike Junkerman, Andy Perkins, Truman Chidsey, Chris Hipes, Bryan Moore, Jeff Black, Tim Watson, and Philip Phibbs of Vulcan Materials Co.

The authors thank the many current and former researchers and technicians with the National Institute for Occupational Safety and Health (NIOSH) who assisted with the implementation of the three ergonomics processes: Bridger Coal Co. Process - Kim C. Gavel, Launa Mallett, Fred Turin, Rich Unger, Charlie Vaught, and William Wiehagen; Badger Mining Corp. Process - Pauline Lewis and Sean Gallagher; and Vulcan Materials Co. Process - Kelly Baron and Susan Moore.

Additionally, several NIOSH researchers participated in the development of the process implementation tools and training described in this document. We extend our appreciation to Jeff Welsh and Jonisha Pollard for assisting with the development of the Risk Factor Cards; Susan Moore for assisting with the development of the Hand Tool Checklist; Bill Porter for graphic modifications to the Risk Factor Checklist and Ergonomics Observations Form; E. William Rossi for graphic support in developing posters and stickers; and Al Cook, Tim Matty, and Mary Ellen Nelson for their assistance in the design, fabrication, and testing of interventions.

ERGONOMICS PROCESSES: IMPLEMENTATION GUIDE AND TOOLS FOR THE MINING INDUSTRY

By Janet Torma-Krajewski, Ph.D.,[1] Lisa J. Steiner,[2] and Robin Burgess-Limerick, Ph.D.[3]

Abstract

Research has shown that an ergonomics process that identifies risk factors, devises solutions to reduce musculoskeletal disorders (MSDs), and evaluates the effectiveness of the solutions can lower worker exposure to risk factors and MSDs and improve productivity. A review of the Mine Safety and Health Administration (MSHA) injury/illness database indicated that 46% of illnesses in 2004 were associated with repetitive trauma and 35% of nonfatal lost days involved material handling during 2001–2004. Even though these statistics show that MSDs significantly contribute to occupational illnesses and injuries in the U.S. mining industry, few mining companies have implemented an ergonomics process. Despite the many unique challenges in the mining environment, three mining companies partnered with the MSD Prevention Team at the National Institute for Occupational Safety and Health's Pittsburgh Research Laboratory to demonstrate that an ergonomics process could be systematically implemented and effectively integrated with existing safety and health programs. Because these three mining companies were very different in organization, culture, and size, the ergonomics processes had to be modified to meet the needs of each company. A description of how these three companies applied ergonomics and the tools and training used to implement their processes is given. Prior to discussing the case studies, general information on the elements of an ergonomics process is provided.

---

[1]Lead Research Scientist, Pittsburgh Research Laboratory, National Institute for Occupational Safety and Health, Pittsburgh, PA.
[2]Team Leader, Musculoskeletal Disorder Prevention Team, Pittsburgh Research Laboratory, National Institute for Occupational Safety and Health, Pittsburgh, PA.
[3]Associate Professor in Occupational Biomechanics, School of Human Movement Studies, The University of Queensland, Brisbane, Australia.

Section I
Introduction: Ergonomics and Risk Management

Ergonomics is the scientific discipline concerned with the understanding of interactions among people and other elements of a system to optimize their well-being and overall system performance [IEA 2008]. This is generally accomplished by applying ergonomic principles to the design and evaluation of manual tasks,[1] jobs, products, environments, and systems, ensuring that they meet the needs, capabilities, and limitations of people. When integrated with safety and health programs, ergonomics can be viewed as a third leg of a three-pronged risk management approach to reduce musculoskeletal disorder (MSD) rates. Safety focuses on hazards that may result in traumatic injuries, industrial hygiene concentrates on hazards that may cause occupational disease, and ergonomics addresses risk factors that may result in MSDs and other conditions, such as vibration-related illnesses. By applying ergonomic principles to the workplace with a systematic process, risk factor exposures are reduced or eliminated. Employees can then work within their abilities and are more efficient at performing and completing tasks. The benefits of applying ergonomic principles are not only reduced MSD rates, but also improved productivity and quality of life for workers.

The purpose of this document is to provide information on implementing a successful ergonomics process that is part of the organizational culture. Section I describes the basic elements of the process and then discusses the importance of employee participation in the implementation of the process. Also included in this section is information on the evolution of risk management as it applies to an ergonomics process. A model developed for safety and health risk management defines five stages, ranging from a pathological stage to a generative stage—from a stage that attributes safety problems to employees to one that involves all employees in risk management at multiple levels with the goal of promoting the well-being of employees. Section II describes how three mining companies implemented ergonomics processes, including lessons learned. Interventions implemented by the mining companies are presented in Section III, along with information on changes to discomfort levels at one of the companies. Section IV describes various tools used when implementing the processes, while Section V focuses on

---

[1]Manual tasks are tasks that involve lifting, pushing, pulling, carrying, moving, manipulating, holding, pounding, or restraining a person, animal, or item.

training, including a presentation for management that promotes the value of ergonomics processes. The tools presented in Section IV and the management presentation contained in Section V are provided as electronic files on the CD included with this document.

Basic Elements of Ergonomics Risk Management Processes

Successful ergonomics risk management processes have several elements in common. The process starts with establishing an understanding of the task and interactions that occur between the worker and equipment, tools, work station used to complete the task, and work area/environment in which the task is conducted. Managing risks associated with manual tasks requires identifying risk factor exposures. If the exposures cannot be eliminated, the degree and source of risk requires assessment. Potential controls or interventions are then identified, evaluated, and implemented to reduce the risk as far as reasonably practical.

> **The ultimate aim of an ergonomics risk management process is to ensure that all tasks performed in workplaces can be performed with dynamic and varied movements of all body regions with low to moderate levels of force, comfortable and varied postures, no exposure to whole-body or hand-arm vibration, and breaks taken at appropriate intervals to allow adequate recovery.**

Element 1: Identifying Risk Factor Exposures During Manual Tasks

Identification of risk factor exposures should include consultation with employees, observation of manual tasks, and/or review of workplace records. Employees should be asked what they think is the most physical part of their job or what task is the hardest to do. Conditions that could potentially indicate risk factor exposures include the following:

- An MSD was associated with performance of the task.
- Any employee is physically incapable of performing the task.
- The task can only be done for a short time before stopping.
- The mass of any object being handled exceeds 35 pounds.
- The postures adopted to perform the task involve substantial deviations from neutral, such as reaching above shoulders, to the side, or over barriers; stooping; kneeling; or looking over shoulder.
- The task involves static postures held for longer than 30 seconds and is performed for more than 30 minutes without a break or for more than 2 hours per shift.
- The task involves repetitive movements of any body part and is performed for more than 30 minutes without a break or for more than 2 hours per shift.
- The task is performed for more than 60 minutes at a time without a break.
- The task is performed for longer than 4 hours per shift.
- Any employee reports discomfort associated with the manual task.
- An employee is observed having difficulty performing the manual task.
- Employees have improvised controls for the task (e.g., phone books for footstools, use of furniture other than that provided for the task).
- The task has a high error rate.
- Workers doing this task have a higher turnover, or rate of sick leave, than elsewhere in the organization.
- Exposure to whole-body vibration (vehicles) or arm-hand vibration (power tools) exceeds 2 hours per shift.

NOTE: The conditions listed above were compiled by the authors based on their professional knowledge and from various sources, such as the Washington State Hazard and Caution Zone Checklists [Washington State Department of Labor and Industries 2008a,b] and limits used for medical restrictions and other guidelines. These conditions alone do not necessarily indicate a risk factor exposure, nor do they indicate a boundary between safe and unsafe conditions. Rather, they must be evaluated in terms of the worker and all aspects of the task: methods or work practices, equipment, tools, work station, environment, duration, and frequency.

If after adequate consultation, observation, and review of records, none of the above conditions is met for any manual tasks in a workplace, then it is reasonable to conclude that the manual tasks are likely to constitute a low MSD risk. For each manual task that has been identified as requiring assessment (one or more of the above conditions is identified), it is sensible to ask whether the task can be easily eliminated. If the manual task can be eliminated, and this is done, then there is no need for an analysis. Reassessment should be conducted whenever there is a change in equipment or work processes. Any new MSD or report of discomfort that is associated with any manual task should trigger either elimination of the task or a risk assessment.

Element 2: Assessing MSD Risks for Manual Tasks

If risk factor exposures exist that cannot be eliminated, the next step is to assess the risks. The aim of the risk assessment is to assist the risk control process by providing information about the root causes and severity of the risk. The assessment should be undertaken with the involvement of the workers who perform the tasks. The assessment of exposures is complicated by the number of exposures that contribute to determining the MSD risk and by the interactions among the different risk factors. The risk assessment process is also complicated by the number of body parts that can be affected and by the variety of possible ways in which an MSD may occur. MSDs occur when the forces on a body tissue (muscle, tendon, ligament, and bone) are greater than the tissue can withstand. MSDs do not occur suddenly as a consequence of a single exposure to a force. They arise gradually as a consequence of repeated or long-duration exposure to lower levels of force. Even low levels of force can cause small amounts of damage to body tissues. This damage is normally repaired before an MSD occurs. However, if the rate of damage is greater than the rate at which repair can occur, an MSD may result. MSDs may also result from a combination of these mechanisms, e.g., a tissue that has been weakened by cumulative damage may be vulnerable to sudden injury at lower forces. Also, if a tissue has suffered a sudden injury, it may be more prone to an MSD-type injury during its recovery process. Manual task risk assessment needs to consider these possible mechanisms. MSDs associated with manual tasks can occur to a range of different parts of the body, and the injury risks associated with a task will vary for different body regions. Consequently, the degree of exposure to different risk factors must be assessed independently for different body regions. In addition to the forces involved, the risk of an MSD to a body part depends on the movements and postures involved, the duration of the exposure, and whether there is exposure to vibration. The risk assessment must address each of these risk factors and the interactions between them.

The first step in assessing the risk of an MSD associated with a particular manual task is to determine the body regions of interest. This may be self-evident if the task has already been identified as causing MSDs or discomfort to a particular body part or parts. Alternatively, the risk assessment should consider the risk of an MSD to each of the following regions independently: lower limbs, back, neck/shoulder, and elbow/wrist/hand. MSDs are most likely to occur when significant exposure to multiple risk factors occurs. Primary risk factors include forceful exertions, awkward postures, static posture, repetition, and vibration. Combining these risk factors greatly increases the risk for developing an MSD. Each of these risk factors is described briefly below.

Forceful Exertions

An important factor in determining the likelihood of an MSD to a specific body part is how much force is involved. Historically, the mass of objects being handled has been the focus. However, the risk associated with a task depends on a number of other factors as well. For example, in lifting and lowering tasks, the force required by the back muscles can depend on the distance of the load from the body as well as the mass of the load. Similarly, if the task involves pushing or pulling a load, the force involved will depend on the frictional properties of the load and the surface, along with the mass of the load.

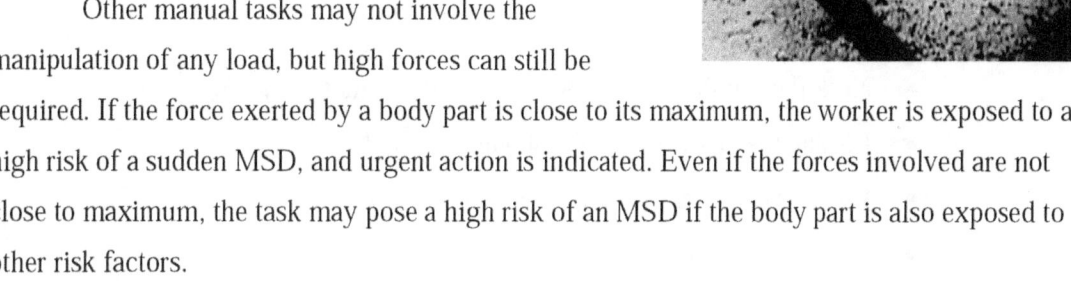

Other manual tasks may not involve the manipulation of any load, but high forces can still be required. If the force exerted by a body part is close to its maximum, the worker is exposed to a high risk of a sudden MSD, and urgent action is indicated. Even if the forces involved are not close to maximum, the task may pose a high risk of an MSD if the body part is also exposed to other risk factors.

High-speed movements (hammering or throwing) are an indication of elevated risk, mostly because high speed implies high acceleration, which in turn implies high force, especially if the speed is achieved or stopped in a short time. Such "jerky" movements are an indication of initial high exertion of the body parts involved. This also includes rapid changes in the direction of movement. Another high-force situation occurs when impact force is applied by the hand to strike an object or surface. In this case, there is a high force applied to the hand by the object or surface being struck.

The magnitude of the force relative to the capabilities of the body part is what is important in assessing MSD risks. For example, the small muscles of the hand and forearm may be injured by relatively small forces, especially if the task is executed at extremes of the range of movement at a joint. This also implies that the capability of the individual performing the work must be taken into consideration when assessing the MSD risk. Overexertion depends on the magnitude of the force relative to the capabilities of the structures.

Awkward Postures

The body postures used during a task influence the likelihood of an MSD in a number of ways. If joints are exposed to postures that involve range of movement near the extreme positions, the tissues around the joint are stretched and the risk of an MSD is increased. Ligaments, in particular, are stretched in extreme postures. If the exposure to extreme postures is prolonged, the ligaments do not immediately return to their resting length afterwards. Tissue compression may also occur with extreme postures. For example, extreme postures of the wrist increases the pressure within the carpal tunnel, resulting in compression of the median nerve as it passes through the carpal tunnel.

The following list provides examples of awkward postures that may involve range of movement near extreme positions [Washington State Department of Labor and Industries 2008a,b; OSHA 1995]:

- Neck flexion (bending neck forward greater than 30°)
- Raising the elbow above the shoulder
- Wrist flexion greater than 30°
- Back flexion greater than 45°
- Squatting

Other joint postures are known to be associated with increased risk of discomfort and MSDs. These include:

- Trunk rotation (twisting)
- Trunk lateral flexion (bending to either side)
- Trunk extension (leaning backward)
- Neck rotation (turning head to either side)
- Neck lateral flexion (bending neck to either side)
- Neck extension (bending neck backwards)
- Wrist extension (with palm facing downward bending the wrist upward)
- Wrist ulnar deviation (with palm facing downward bending the wrist outward)
- Forearm rotation (rotating the forearm or resisting rotation from a tool)
- Kneeling

There are other awkward postures that increase the risk of an MSD because of the orientation of the body with respect to gravity and do not necessarily involve extreme ranges of movement. These postures usually require the worker to support the weight of a body part. An example would be lying under a vehicle to complete a repair. When assessing postures, it is important to note that workers of different sizes may adopt very different postures to perform the same task.

The force exertion of muscles is also influenced by the posture of the joints over which they cross. Muscles are generally weaker when they are shortened or lengthened. This effect will be greatest when the joints approach the extremes of the range of movement. Consequently, the optimal design of work aims to provide tasks that can be performed while maintaining neutral postures. The following are descriptions of neutral postures for different body parts [OSHA 2008; Warren and Morse 2008]:

Head and neck	Level or bent slightly forward, forward-facing, balanced and in line with torso
Hands, wrists, and forearms	All are straight and in line
Elbows	Close to the body and bent 90° to 120°
Shoulders	Relaxed and upper arms hang normally at the side of the body
Thighs and hips	Parallel to the floor when sitting; perpendicular to the floor when standing
Knees	Same height as the hips with feet slightly forward when sitting; aligned with hips and ankles when standing
Back	Vertical or leaning back slightly with lumbar support when sitting; vertical with an S-curve when standing

Static Posture

The optimal design of work results in tasks that involve slow to moderately paced movements and varied patterns of movement. Little or no movement at a body part elevates the risk of discomfort and MSDs because the flow of blood through muscles to provide energy and remove waste depends on movement. Tasks that involve static postures quickly lead to discomfort, especially if combined with exposure to other risk factors.

Repetition

If the task involves repetitively performing similar patterns of movement, and especially if the cycle time of the repeated movement is short, then the same tissues are being loaded in the same way with little opportunity for recovery. Such repetitive tasks are likely to pose a high risk of cumulative injury, especially if combined with moderate to high forces (or speeds), awkward postures, and/or long durations.

Vibration

Exposure to vibration in manual tasks comprises two distinct types: hand-arm vibration (typically associated with power tools) and whole-body vibration (typically associated with vehicles). In both cases, the vibration exposure impacts MSD risk both directly and indirectly.

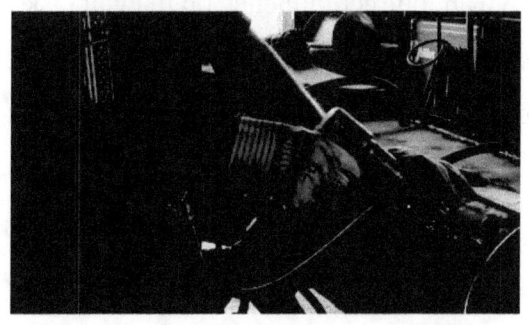

Exposure of the upper limbs, and particularly the hands, to high-frequency vibration associated with power tools is a direct cause of damage to nerves and blood vessels. Short-term effects are temporary loss of sensation and control, and blanching of the fingers (vibration white finger syndrome). These effects may become irreversible with long-term exposure and lead to gangrene and loss of the affected fingers [NIOSH 1989]. Use of vibrating power tools is also an indirect cause of MSD risk to the upper limbs because the vibration increases the force required by the upper limbs to perform the task. The degree of risk increases with higher-amplitude vibration tools (hammer drills or jackhammers).

Similarly, long-term exposure to whole-body vibration (typically from vehicles) is associated with back pain [Bovenzi and Hulshof 1999; Lings and Leboeuf-Yde 2000; ACGIH 2007a]. As well as a direct effect on the back, exposure to whole-body vibration also has an indirect influence on MSD risk by causing fatigue of the back muscles. Again, the risk is greater when the amplitude of vibration is high (heavy vehicles and/or rough terrain).

Another important consideration is the duration of the exposure. If a task is performed continuously, without a break and for a long time, the tissues involved do not have opportunity for recovery, and the risk for a cumulative injury increases. Performing several tasks during a shift can provide recovery if the tasks involve different body parts and movement patterns.

In general, a root cause is defined as a source of a problem. In terms of MSD risk factor exposures, it is important to determine why the exposure is occurring or to identify the root cause of the exposure. Root causes modify the degree of risk in two ways. Some root causes are characteristics of the work that commonly lead to increased exposure to the risk factors discussed previously. Modification of these root causes will likely reduce the MSD risk. Other root causes have an indirect influence on manual task MSD risk. Understanding the root causes of risk factor exposures can help determine the most effective means for reducing or eliminating the exposures. Examples of root causes include the following:

Workplace or Work Station Layout
- Working in confined spaces is likely to result in the necessity to adopt awkward postures to perform tasks.
- Work stations with restricted visibility typically result in awkward and static postures, especially of the neck.
- Work stations with inappropriate location of visual displays (usually too high or located to one side) cause awkward postures, especially of the neck.
- Standing work leads to fatigue if undertaken for long durations.
- Kneeling work causes high force on the knees.
- Working below the height of the feet inevitably leads to extreme trunk postures.
- Working overhead requires awkward and static postures of the shoulders.
- Work stations that require reaching to handle objects create awkward postures.
- Work surfaces that are too high or too low lead to awkward postures.
- Locating objects to be handled below knee height results in trunk flexion.
- Locating objects to be handled above shoulder height leads to working with the elbows above the shoulders.
- Carrying loads for long distances results in fatigue.

Objects, Equipment, and Tools

- Any unpredictability, such as handling an object with uneven or shifting distribution of its mass, may lead to overexertion of muscles.
- Handling heavy loads, even if they are not lifted, may require high force because of the inertia of the load.
- Handling large loads, even if they are not heavy, may require high forces because of the distance of the center of the load from the body.
- Objects that are hot, cold, or otherwise noxious may lead to the load being held away from the body, which increases stress on the lower back and shoulders.
- Objects with handles may result in contact stress or decreased control of the object.
- Poorly maintained tools (i.e., dull bits or blades) may increase the force required.
- Using tools not appropriate to the task (too powerful or not powerful enough, too heavy, incorrect handle orientation, etc.) may lead to awkward postures and forceful exertions.
- Handling loads with one hand results in only one side of the body supporting the load, which could lead to overexertion.
- Triggers that require sustained force or are operated with a single finger may lead to fatigue and overexertion.
- Gloves generally increase the force requirements of a task.

Environmental Conditions

- Low lighting levels or glare may cause awkward postures or prolonged squinting of the eyes.
- Exposure to hot environments increases fatigue, especially for heavy work.
- Exposure to cold, in addition to other risk factors, is implicated in the development of vibration white finger syndrome or hand-arm vibration syndrome, and carpal tunnel syndrome from increased hand forces generated as a result of wearing gloves and cold hands.

- Uneven or poorly maintained surfaces can increase forces required to push/pull carts, the amplitude of whole-body vibration, or the likelihood of slips and falls.

Work Organization and Systems

Certain factors of work organization and systems may lead to fatigue and overexertion of muscle groups. In some cases, recovery times do not permit the worker to return to baseline values prior to returning to work. Examples of such factors include:

- High work rates
- Lack of task variety
- Uneven temporal distribution of work causing high peak loads
- Understaffing
- Irregular or long shifts
- Pay schemes that encourage working faster or longer

Studies have shown that even when controlling for higher workloads, elevated rates of discomfort and/or MSDs still occurred because of the presence of other work organization and system factors not typically associated with discomfort or MSDs [Bernard 1997]. The physiological mechanism for this effect is not well understood. Addressing these factors in addition to implementing controls that reduce risk from higher workloads may increase success at reducing rates of discomfort or MSDs. These factors may include:

- Job dissatisfaction
- Perception of intensified workload
- Lack of job control
- Uncertainty about job expectations
- Lack of opportunity for communication and personal contact
- Cognitive overload, monotonous work, frequent deadlines, interpersonal conflict

Element 3: Controlling MSD Risks During Manual Tasks

There are several ways to reduce MSD risks that occur during the performance of manual tasks. From an ergonomics perspective, the emphasis is first on eliminating or reducing risk through design controls; secondly on administrative controls, such as job rotation or enlargement; and then on personal protective equipment (PPE). When risks cannot be eliminated with design controls, administrative controls and PPE may also be required to manage the residual risks. Regardless of which controls are chosen, training is an important aspect of the implementation to ensure that workers are aware of the appropriate way of performing work and using equipment.

Elimination

Having determined that manual tasks with risk factor exposures are performed in a workplace, the next step is to determine whether any or all of the manual tasks can be eliminated. If this is possible, it is the most effective way of reducing MSDs. Some manual tasks can be eliminated by examining the flow of materials and reducing double handling. Others may be eliminated by changing to bulk-handling systems. Outsourcing manual tasks may also be considered as a way of eliminating exposures to your workers if the organization undertaking the task has specialized equipment that reduces the risk for its workers to acceptable levels. It would not be appropriate to outsource manual tasks if the risk was not reduced. Some tasks, such as cleaning up waste, are nonproductive and may be eliminated or reduced by examining the source of the waste.

Design Controls

If, after the possibilities have been examined, it is determined that some hazardous manual tasks cannot practically be eliminated, and the risks associated with these tasks have been assessed, the next step is to devise design controls that will reduce the MSD risks. This step is most effectively undertaken in consultation with all workers who will be affected by the change, including maintenance as well as operational staff. Apart from the fact that workers are the ones who know most about the tasks, the probability of success of the design changes is enhanced if the workers concerned have a sense of ownership of the changes. Before implementing the design controls, it is also important to consider whether new hazards will be introduced as a consequence of the control.

Considering the following aspects of the work area and task is a useful way of thinking about possible design controls:

Work Areas: Work Height, Space, Reach Distances, Work Flow, Adjustability

The design of work areas has a large impact on MSD risks. For example, limited space, limited clearances, and restricted access to work are common causes of awkward postures. Work should be located at an appropriate height and close to the body. Providing adjustability of work stations may be an option to accommodate workers of different sizes. Workplaces should be designed to increase postural variability during work.

Loads: Size, Shape, Weight, Stability, Location, Height

The nature of loads that are delivered to a workplace, handled within a workplace, or produced by a workplace are a common source of risk factor exposures when performing manual tasks. Increasing the size and mass of loads and implementing mechanized bulk-handling systems are effective design controls. Reducing the size and weight of loads is another option, but may require training and ongoing supervision to ensure that multiple loads are not handled simultaneously to increase speed. Ensuring loads are easily gripped by providing or incorporating handles is important. Hot or cold loads should be insulated, or proper protective clothing should be provided to allow the loads to be comfortably held close to the body. Where loads are manually handled, they should be stored at waist height rather than on the floor or above shoulder height.

Tools: Size, Weight, Handles, Grips, Trigger, Vibration

Poorly designed handtools are a common source of awkward postures, high exertion (particularly of the small muscles of the hand and arm), and hand-arm vibration. Handtools should be designed such that joint postures remain close to neutral during use and should be as light as possible. Heavy tools may be supported by a counterbalance to reduce exertion. While power tools reduce exertion, the vibration associated with power tools introduces a new risk, and tools and consumables should be chosen to minimize the amplitude of the vibration as far as possible. Tools also need to be maintained (e.g., keep blades and bits sharp) to minimize vibration levels.

Mechanical Aids: Hoists, Overhead Cranes, Vacuum Lifters, Trolleys, Conveyers, Turntables, Monorails, Adjustable Height Pallets, Forklifts, Pallet Movers

A large number of different mechanical aids are available to reduce risk factor exposures, and these can be effective controls. However, care is required to ensure that the use of the aid does not significantly increase work performance time. If it does, the likelihood that the control will be effective is reduced because administrative controls and ongoing supervision will be required to ensure use. Introducing mechanical equipment, such as forklifts, also introduces new risks that require control. For example, using forklifts requires that traffic patterns be established and visual obstructions be eliminated.

The design of mechanical aids requires careful consideration. For example, cart wheels should be as large as possible to reduce resistance (getting stuck in cracks), and vertical handles should be provided that allow the cart to be gripped at different heights by different sized workers. Where mechanical aids are introduced to control manual tasks risks, it is important to ensure that they are maintained in working order and are available when and where required.

Further information on mechanical aids can be found in *Ergonomic Guidelines for Manual Material Handling* [NIOSH et al. 2007].

Administrative Controls

For situations where there are no effective design controls or the design controls that are implemented do not fully address the exposures, it may also be necessary to consider additional administrative controls. Administrative controls rely on human behavior and supervision and, on their own, are not an effective way of controlling manual task MSD risk. Administrative controls include the following:

Maintenance

Maintenance of tools, equipment, and mechanical aids is crucial, but requires a schedule to be developed and supervision to ensure that it occurs. Following a regular schedule of preventive maintenance not only impacts productivity, but can also reduce exposures to risk factors. For example, preventive maintenance for mobile equipment can avoid major repair tasks that usually involve exposures to several risk factors, such as excessive force, awkward postures, and vibration. Another aspect of maintenance is good housekeeping.

Workload

MSD risk associated with manual tasks may be reduced by reducing shift duration or the pace of work. It may be possible to change the distribution of work across the workday or week to avoid high peak workloads. Ensuring that appropriate staffing levels are maintained is important. Provision of adequate rest breaks can reduce MSD risks.

Job Rotation and Task Variety

It may be possible to reduce MSD risks by rotating staff between different tasks to increase task variety. This requires that the tasks are sufficiently different to ensure that different body parts are loaded in different ways. Alternatively, multiple tasks might be combined to increase task variety.

Team Lifting

Team lifting may be effective in reducing injury risk where the load is bulky, but relatively light. However, if the load is not "heavy enough," an employee may try to handle the load individually, especially if there are not many other employees in the area. If team lifting is used as a control, training and supervision are required to ensure that the task is only done when appropriate staff are available to perform the task.

Personal Protective Equipment

Some forms of PPE may be effective in reducing risk factor exposures. However, PPE only serves as a barrier, and the protection provided depends on the effectiveness of the barrier. Consequently, PPE should only be used when risk factor exposures cannot be eliminated or effectively reduced with design controls, or design controls are not economically feasible. PPE may also be considered as an interim control when design controls cannot be implemented in a timely manner. Kneepads, protective aprons, cooling garments, and antivibration gloves are examples of PPE.

Element 4: Monitor and Review

Managing manual task risk is an iterative "continuous improvement" process. Following implementation of any control measure, it is important to check that the controls are working as anticipated and that new risks have not been introduced. It is important to evaluate the effects on not just the workers directly involved with the change, but also other workers and processes that may be affected. Although this element is critical to successful processes, it is sometimes ignored or forgotten as the next issue or problem that arises usually needs the same resources to resolve.

Element 5: Record-Keeping

Keeping records of the steps taken in the risk management process is important for several reasons. It will ensure that an effective risk management process is in place by documenting the changes in risk factor exposures and MSD incident/severity rates. It provides a way of tracking the improvements made, maintaining the corporate memory of the reasons that changes were made, and allows for justification of future changes. Documenting controls or task improvements also allows this information to be shared so that similar tasks at other sites may also be improved using the same or similar controls.

Participatory Ergonomics

"Participative ergonomics" is based on an underlying assumption that the workers involved are the "experts" and must be involved at each stage of the risk management process if it is to be successful. In an MSD management context, employees and management participate jointly in hazard identification, risk assessment, risk control, and evaluation of the risk management process.

Many variations in the models and techniques used in participative ergonomics have been developed [Haines and Wilson 1998; Haims and Carayon 1998; Laing et al., 2005; Burgess-Limerick et al. 2007]. However, a common element is to ensure the use of expert knowledge that workers have of their own tasks by involving the workers in improving their workplaces. Management commitment and provision of resources including a champion to promote the process, workers' and management understanding of relevant ergonomics concepts and techniques, and a process to efficiently develop and implement suggested controls are also important components of successful participative ergonomics interventions.

Using participative ergonomics to address MSDs associated with manual tasks usually entails an ergonomics team, which includes workers as team members. This team must be knowledgeable about the risk management process, have the skills and tools required to assess manual task risks, understand the risk control hierarchy, and have knowledge of general principles of control strategies for eliminating and controlling manual task risks. Implementing an effective ergonomics risk management process also requires that all employees be able to identify risk factor exposures associated with manual tasks and be aware of the aspects of manual tasks that increase MSD risks. Having this awareness allows employees to consider ways to improve their jobs and ultimately reduce risk factor exposures. Training in risk assessment and control strategies ensures successful participation of workers in an ergonomics risk management process. Training team members to acquire these skills and work within a risk management process is a key concern. Team members identify risk factor exposures associated with their work and follow a risk assessment process that develops control suggestions. The team members plan the implementation of key controls and are subsequently shown how to evaluate those controls. Management commitment and effective risk management systems are required in order for the approach to be effective. Access to external ergonomics expert assistance may be necessary for particularly difficult or complex problems. It is also important to note that ergonomics is equally concerned with improving productivity and reducing waste, as well as reducing injury risks [Dul 2003]. This is crucial because any work modification that is implemented to reduce MSD risk should be easier, quicker, or more efficient than the previous methods of work. If not, the chance of acceptance and adherence to the new methods is markedly reduced, and ongoing supervision will be required to ensure compliance.

Evolution of Risk Management Processes

A risk management model, originally developed by Westrum [1991] and Westrum and Adamski [1999] and later broadened by Hudson [2003], describes the evolution of risk management strategies and the progression as a company moves from a pathological to a generative stage with regard to how risk is managed (Figure 1). At one end of the spectrum, the *pathological* stage can be thought of as the stage in which safety problems are attributable to the workers. The main driving force is the business and not getting caught by regulators. The *reactive* stage is the point where companies consider safety seriously, but only intervene following the occurrence of accidents. At the *calculative* stage, safety is driven by management systems; it is still imposed by management and not sought by the workforce. In the *proactive* stage, the workforce is becoming increasingly active in risk management. Finally, in a *generative* stage, everyone is involved in risk management and tries to maintain the well-being of themselves as well as their coworkers.

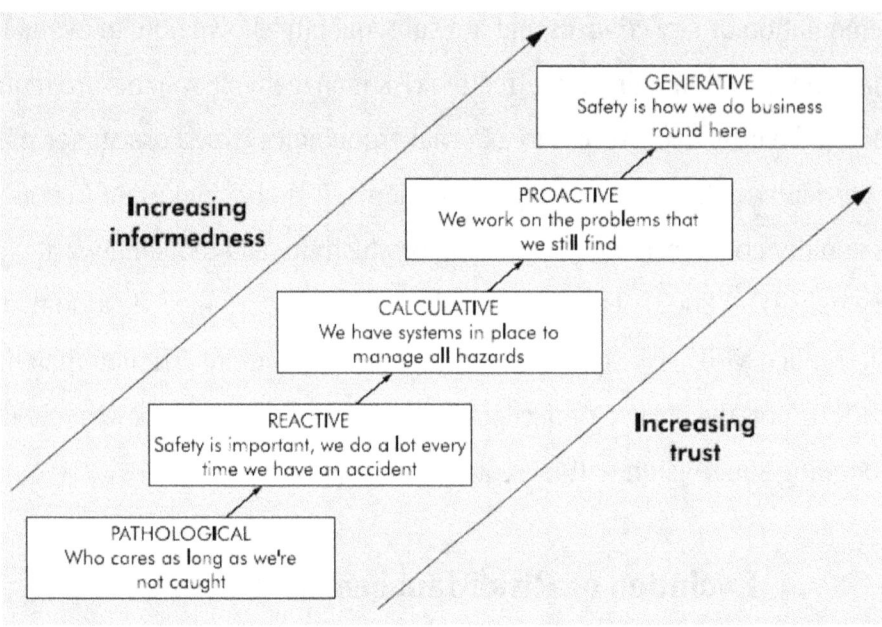

Figure 1.—Evolution of health, safety, and environment risk management process [Hudson 2003].

This risk management hierarchy may be applied to an ergonomics risk management process where the company and the workforce integrate ergonomics principles into their risk management process. In this case, the approach follows the same path but with a focus on eliminating MSDs.

Pathological Stage: Workers and companies are unaware of how MSDs occur and let workers look out for themselves. Employees may have the signs and symptoms of an impending MSD, but no changes are made to the workplace. No formal job safety analysis techniques are used, and productivity is the primary focus.

Reactive Stage: Analysis of the incident is after the report of an MSD or several MSDs, and the solution or correction is often individualistic. Others doing similar jobs may or may not be considered as it is thought to be one particular employee's problem. For MSD-related issues, often the workers believe that aches and pains are just part of their jobs or the aging process. They do not know that these recurring aches and pains are precursors to cumulative injuries and that these injuries can be prevented through planning of jobs, work environment, and equipment purchasing.

Calculative Stage: At this stage, companies may accuse workers of being "hurt at home" or by "their hobbies" rather than by their work environment or by poor work task design or planning. Some management may use some outside training for proper lifting techniques or purchase "ergonomically designed" PPE or equipment to resolve issues. In some cases, the company may fix very specific problems successfully through training and procedural approaches. These interventions have a positive impact on the situation, but the more global philosophy of prevention is not adopted. In addition, there is no formal followup to see if the problem was resolved or if any other problems have resulted. In this stage, management may be aware of the cumulative injury process, but employees are not. Safety is still in the hands of management and not pushed down to the employee level. Management believes that the system in place works well to address issues brought to their attention.

Proactive Stage: Employees are educated about ergonomics principles, cumulative injury progression, and techniques to identify and reduce risk factors associated with MSDs. Management relies on employees to bring issues to them and to resolve them together. Management may also seek to provide periodic observations of all tasks or establish a wellness or fit-for-duty program. Ergonomic principles are used when evaluating and redesigning jobs. Management and workers are not waiting for MSDs to occur, but rather are looking for exposures to indicators (risk factors) that point to a potential MSD and then reduce or eliminate that exposure. In some cases, a consultant in ergonomics may be hired or an ergonomics committee formed. Focusing on risk factor exposures and reports of MSDs investigates *why* (root causes) such situations are occurring instead of *what* or *when*. The company takes responsibility for employees' health during and outside of work and places less blame on the employee. Job safety analysis techniques include the evaluation of risk factors at each step in the standard operating procedures to ensure that they are considered. Finally, a procedure is put in place to conduct followup that ensures the solutions worked and to investigate other emerging issues. Anecdotally, workers appreciate these analyses and believe it is in their own interest and not just the company's interest. Most solutions are off-the-shelf, and lessons learned are communicated throughout the mine and even company-wide. Still, the value (cost/benefit) of these interventions may not be fully understood and consequently may be underreported.

Generative Stage: There is anticipation of issues with regard to old and new processes and equipment. The ergonomic principles are integrated into the designing and planning processes. This integration occurs in the beginning and is understood to be as important as other engineering and purchasing decisions. Employees are trusted to make decisions about their jobs and recognize situations where changes need to be made. At this point, the employees are empowered with resources to make changes and inform management of needs. Investigation of risk factors, signs, and symptoms of MSDs is driven by an understanding of their root causes. The solutions are cost-effective and creative, and followups are done automatically. A database of all reported issues and changes to the workplace and equipment is available to the entire company and serves as an informational base from which to make the best purchasing and planning decisions. Safety is in the hands of educated employees. The cost of MSDs or cumulative injuries is reduced and profits are increased, the workforce returns home healthy, operating procedures include ergonomic principles, better habits are passed on to new recruits, and management and employees together see the overall interaction of systems and people. Less time is spent on addressing health and safety issues because they are under control and are the responsibility of all parties.

There are many characteristics of these stages not addressed here. However, the above is a summary of what a company might expect as it moves toward a more generative risk management approach. A company can use these descriptions to measure where they are and how to get to where they want to be [Shell International 2003]. The first step to achieving generative status is to understand what information is needed and how to educate employees to help themselves and their coworkers.

Section II
Ergonomics Processes: Case Studies

Mining is often characterized by physically demanding tasks performed under dynamic conditions, which creates greater challenges for applying ergonomic principles [Steiner et al. 1999; Scharf et al. 2001]. To demonstrate the efficacy of applying ergonomic principles in mining environments, the National Institute for Occupational Safety and Health (NIOSH) partnered during 2000–2007 with three mining companies, different in size, organizational structure, and culture. Descriptive information about each company is provided in Table 1.

Table 1.—Demographic information for the three mining companies that partnered with NIOSH to implement ergonomics processes at their mines

	Mining company		
	Bridger	**Badger**	**Vulcan**
Company size	1 mine	2 mines	372 facilities (175 mines)
Location	Wyoming	Wisconsin	21 states
Type of mine	Surface	Surface	Surface
Commodity	Coal	Sandstone	Gravel
Mining process	Drill-blast-dragline/dozer-drill-blast-load-haul	Drill-blast-load-haul and sand-water slurry pumped to processing plant	Drill-blast-load-haul
No. of employees	350	180	8,000 plus – usually fewer than 50 employees at each pilot site
Unionized workforce?	Western Energy Workers Union	No	No
Safety program	Safety Department and Safety Committee	Safety Team	Safety, Health, and Environmental (SHE) Team and division- and corporate-level support
Behavior-based Safety System	No	Yes	No

All three companies embraced the process elements described in Section I and identified by Cohen et al. [1997], but how these elements were addressed varied. This section illustrates how the three mining companies applied ergonomic principles and adapted the implementation process to meet their organizational and cultural needs. Tools and training used during the implementation of these processes are described in later sections.

Bridger Coal Co.

The first mine that NIOSH worked with was the Jim Bridger Mine, a surface coal mine located 35 miles northeast of Rock Springs, Sweetwater County, WY. This mine had one active pit approximately 20 miles long and an average production rate of 6.4 million tons of coal per year during 1995–2000. The workforce comprised 350 employees. The mine was operated by the Bridger Coal Co., a PacifiCorp company and subsidiary of Scottish Power.

For 5 years prior to this project, the average incidence rate for nonfatal days lost (NFDL) injuries at the Jim Bridger Mine was 1.32 injuries per 100 employees, compared to the national average of 2.34 for all mines and 1.31 for all western U.S. surface coal mines with more than 100 employees. Although the mine's average incidence rate was well below the national average and injuries related to MSD risk factors did not seem to be a major issue, Bridger Coal Co. decided to implement an ergonomics process. This action was consistent with mine management's proactive approach to safety and health and its culture of seeking continuous improvement.

The Jim Bridger Mine has a very traditional approach to safety and health. This program is managed by a Safety Department and supported by a Safety Committee, with members from several other departments, such as production, maintenance, medical and engineering. Employees were empowered to identify hazards and to request corrective action through their supervisors and/or the Safety Department.

Bridger Coal's management decided that the best approach to implementing an ergonomics process was to establish an Ergonomics Committee within the Safety Department, but separate from the existing Safety Committee. This approach allowed Bridger to commit resources specific to ergonomic interventions. The committee, chaired by an Ergonomics Coordinator who reported to the Safety Manager, included 11 representatives from labor and management. Specific departments represented were medical, engineering/environmental, safety, human resources, production, and maintenance. Mine management was kept informed of committee activities and resource needs through the Ergonomics Coordinator and Safety

Manager, who reported to the Mine Manager. The union was kept abreast of committee actions by union representatives appointed to the committee. The Ergonomics Coordinator and Safety Manager served as champions for the process and ensured that the process moved forward.

Since the Bridger Coal Co. decided to implement its ergonomics process separate but within its safety and health program, it was necessary for the Ergonomics Committee to define a procedure for processing concerns. The committee designed two forms for employees to complete to present concerns for followup: an "Employee Ergonomic Concern" form and a "Risk Factor Report Card." The Employee Ergonomic Concern form requested specific information about equipment and work area, the nature of the concern, and whether the concern was acute or cumulative in nature. The Report Card was a 4- by 6-inch card that gave employees a mechanism to also identify potential risk factors and affected body parts, and note any comments and/or suggestions. Employees could complete either form, or both. The committee encouraged completing both forms since different information was collected by each form.

The steps followed by the Ergonomics Committee for processing a concern are shown in Figure 2. The concern is screened by the committee chairperson to determine if the problem involves exposure to MSD risk factors and if the exposure could be easily controlled without involvement of the committee. If the exposure cannot be resolved immediately, the concern is discussed at the next meeting and then assigned to a committee member for further review, which includes discussions with the employee submitting the concern. Subcommittees investigating concerns usually involve employees directly affected by the concern. When a concern is not considered viable or an intervention is not possible, the concern is reviewed again later as additional information or options, such as new technology, become available to resolve the concern. Concerns and the status of the concerns are maintained in an electronic spreadsheet.

One of the first actions taken to move the ergonomics process forward was to help the committee gain an understanding of ergonomics. The committee received training on the principles of ergonomics, risk factor identification, job prioritization, intervention recommendations, and cost/benefit analysis. During followup training sessions, the committee received instructions on using tools to document interventions, task analyses, and interviews; conducting interviews; videotaping/photographing tasks; and prioritizing interventions. This training, which was conducted by NIOSH personnel, was a combination of classroom instruction and field exercises so members could gain experience in conducting task analyses and identifying risk factors.

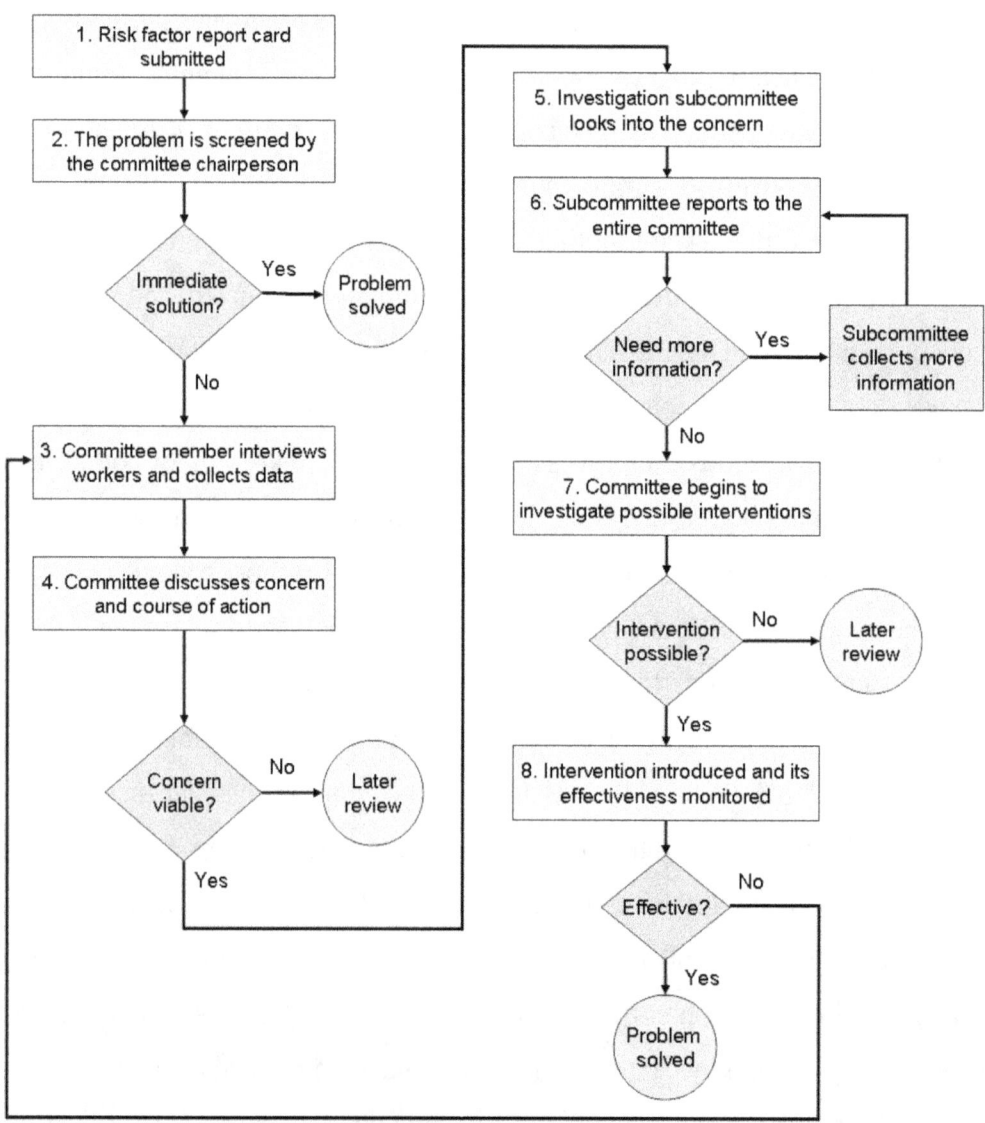

Figure 2.—Flow diagram of Bridger Coal's ergonomics process.

Once the Ergonomics Committee was trained and had developed the procedure for processing concerns, employees were given training that focused on recognizing ergonomic risk factors and taking action by reporting risk factors to the Ergonomics Committee. Employees were told to be proactive and to target risk factors and not wait until an injury occurred. The employees were given information on how a cumulative trauma disorder may develop and how it is better to take action by eliminating risk factors before a disorder occurs. Employees were taught how to report a concern using the Risk Factor Report Card. The primary training module was geared to employees in production and maintenance. A second version of the training focused on office ergonomics and was given to administrative support employees. This 90-minute training was presented by NIOSH personnel and committee members, who introduced the training and then ended the training by encouraging employees to get involved in the process. Approximately 280 employees were trained during 21 sessions. For the most part, the training was well received by the employees. They participated in the interactive exercises and seemed quite knowledgeable about identifying risk factors at the conclusion of the training. In fact, 27 employees submitted Risk Factor Report Cards to the Ergonomics Committee immediately following the training.

A simple record-keeping system was used for the ergonomics process. A listing of concerns was maintained in an electronic spreadsheet that included all the information provided on the Risk Factor Report Card. Additionally, each concern was color coded to document the status of the concern. Concerns were labeled as either completed, in progress, items referred elsewhere or dismissed, or items on hold. The committee also maintained a status/update document that allowed employees to monitor the status of their concerns. This document, posted on the ergonomics bulletin board, provided a short description of the concern and the current status of the intervention. If a concern was referred elsewhere or dismissed, the basis for this decision was provided.

The Ergonomics Committee established a bulletin board in the ready room, an area that all employees passed through when reporting to work. The bulletin board included information about the committee, how to report a concern, and a status report of interventions completed by the committee. NIOSH periodically provided posters to display on the bulletin board and at other meeting areas at the mine. The posters focused on introducing the Ergonomics Committee to the employees, identifying and reporting risk factors, ergonomic interventions completed by the committee, and risk profiles for specific tasks. The posters encouraged participation in the process and promoted interventions. PacifiCorp's quarterly safety newsletter, *Safety Times*, twice featured the success of Bridger Coal's ergonomics process. This newsletter is distributed to all employees of PacifiCorp, including Bridger Coal employees. These articles served as recognition not only to committee members, but also to those employees submitting concerns for actively participating in the process.

The training provided to the committee members and the employees permitted Bridger Coal initially to have a proactive approach to resolving risk factor exposures before an injury or illness occurred. Additionally, employees actively participated in improving their own job tasks. As the process matured, ergonomic principles were applied to other processes, such as equipment purchasing decisions, which moved the ergonomics process to an even higher level of risk management. Because purchase specifications ensured that ergonomic principles were addressed during the construction of the equipment, the equipment arrived at the mine without issues related to risk factor exposures. In just 3 years, the Bridger Coal Co. implemented an effective, proactive process to reduce exposure to MSD risk factors. Instead of waiting for an injury to occur, Bridger Coal relies on an employee-based participative process to implement job improvements that promote the well-being and comfort of its employees and to incorporate ergonomics into many other processes affecting employee safety and health.

> "Ergonomics has played an important role in helping Bridger Coal reach our goal of providing the safest and healthiest working environment possible for our employees. Our management and hourly employees alike understand the value of what has been developed. In the beginning, when the idea of establishing such a program surfaced, we were all skeptical of just how things would work. However, thanks to the combined efforts of NIOSH, PacifiCorp, and those at Bridger Coal Company involved in the creation process, we found that an Ergonomics Program could not only be efficiently developed, but that it could be highly effective as well. The Ergonomics Program is currently an integral part of our company, and we are confident that it will continue to improve and enhance the safe working experience at our mine."
>
> —*Kean Johnson, Ergonomics Process Coordinator*
> *Bridger Coal Co.*

Badger Mining Corp.

Badger Mining Corp. is a family-owned small business with headquarters in Berlin, WI. Badger operates two sandstone mines near Fairwater and Taylor, WI, which produce approximately 2 million tons of industrial silica sand annually. Badger also owns three subsidiary companies, one of which participated in the ergonomics process. This subsidiary (LogicHaul) is located at the Fairwater Mine and is responsible for transportation and distribution of products via trucks and railcars. There are 180 employees at the Resource Center (headquarters offices), Fairwater, Taylor, and LogicHaul.

During 2002–2004, the average NFDL injury incidence rate reported to the Mine Safety and Health Administration was 3.28 injuries per 100 employees for the Taylor Mine. The Fairwater Mine had no NFDL injuries during this period. The national average NFDL injury incidence rate for similar type mines (surface mines that mine the same type of commodity) was 2.15. A review of both NFDL and no days lost (NDL) or restricted workday cases occurring during 2003–2004 at both sites indicated that 79% of the NFDL injuries (61 of 77) and 85% of the NDL injuries (92 of 108) were associated with MSDs.

Organizationally, Badger uses a team management structure consisting of work teams and cross-functional teams that are responsible for setting the work schedule, changing work practices, and providing feedback to the Operations Team. Members of work teams are cross-trained and may perform many disparate tasks. Work teams are self-directed and are responsible for the safety of their members. Badger associates complete CARE (Corrective Action Request for Evaluation) reports for all safety incidents, including accidents, injuries, property damage, near-misses, and hazard exposures. Cross-functional teams address functions pertinent to many teams, such as safety and quality. Each site has a separate Safety Team, which processes the CARE reports and addresses safety-related issues that cannot be resolved by the work teams. Because the mining processes and products are different at the two mines, the members of the two Safety Teams differ slightly. The Fairwater Safety Team includes 25 members and represents 16 work teams; the Taylor Safety Team includes 28 members and represents 15 work teams. The Safety Associate, a headquarters employee, also serves as a member of the Safety Teams at both mines. The Safety Associate functions as a consultant to the mines and provides training, offers motivational programs, conducts investigations, and implements Badger's behavior-based safety (BBS) system, which was initiated in December 2002. BBS observers have been trained to conduct random, periodic observations of employees to identify both safe and unsafe behaviors and to correct unsafe behaviors. Safety observations are documented using a "Do It Safely" form and are conducted at both mines and the Resource Center.

When integrated with safety and health programs, ergonomics can be viewed as an approach to improve injury and illness rates and the overall working conditions for employees by addressing risk factor exposures that may occur during manual tasks. These exposures are most often associated with MSDs, but may also result in other disorders and illnesses, such as heat stress disorders or vibration-related illnesses. Because Badger decided to fully integrate the application of ergonomic principles with its existing safety program, ergonomic concerns are addressed using the same process as any other safety and health concern (see Figure 3). Actions to address these concerns are initiated by either a CARE report or a BBS ergonomic observation, which are reviewed by the Safety Team. If the risk factor exposure(s) can be addressed by this team, then no further action is needed. However, if the cost of the corrective action exceeds the limits set for the Safety Team, then the concern is transferred to the Operations Team. Since the Safety Team includes members of the Operations Team, this transfer is seamless. The champion for the Badger ergonomics process is the Safety Associate.

With a decentralized safety and health process, Badger initiated its ergonomics process by training all employees in February 2005. The training, which lasted 2.5 hours, was given by NIOSH. It emphasized identifying risk factor exposures and then reporting those exposures using a CARE report so that corrective actions could be instituted to resolve the exposures. This training also included a brief introduction to ergonomics and MSDs, with specific information on back injuries and how the risk of injury could change based on methods used to perform lifting tasks. Examples of risk factor exposures were illustrated with short videos of tasks performed at either Badger mine. Training techniques included interactive exercises and demonstrations. To ensure the participation of new associates in the ergonomics process, Badger provides ergonomics and risk factor awareness training during new associate orientation, and to keep associates involved in the ergonomics process, interactive exercises demonstrating ergonomics principles are included in annual refresher training.

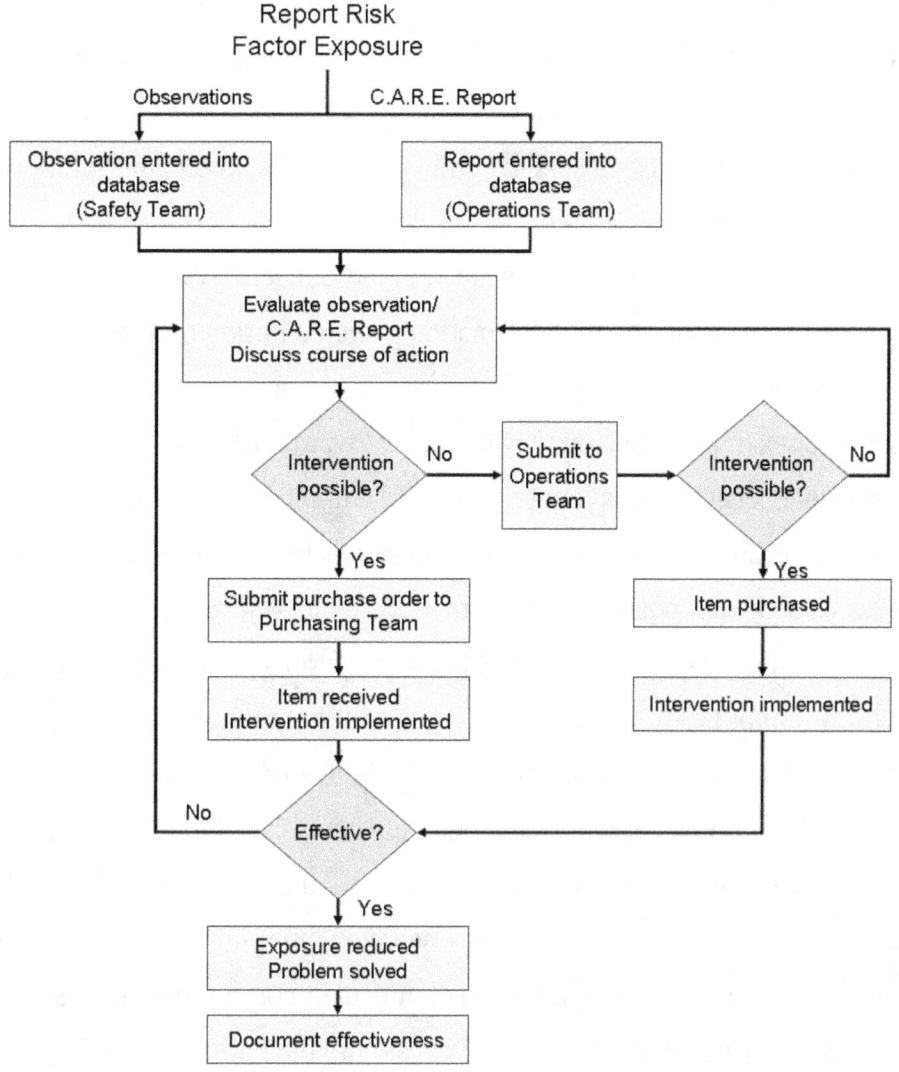

Figure 3.—Flow diagram of Badger company task improvement process.

Because Badger uses a BBS system as part of its overall safety and health program, it was decided to also incorporate ergonomic observations into this system for the purpose of identifying and eliminating exposures to risk factors. The primary focus of a BBS system is to decrease injury rates by preventing unsafe behaviors, which is accomplished by implementing a systematic process of data collection and correction of unsafe behaviors [Krause 2002]. Sulzer-Azaroff and Austin [2000], who examined articles describing the results of implementing BBS systems, reported that 32 of 33 BBS systems reviewed resulted in injury reductions. However, none of these systems reported results specific to MSDs. Although the top three U.S. automakers do not integrate their ergonomics processes with their BBS systems, other automotive companies—Toyota and Tenneco Automotive—have done so. In these two companies, BBS systems were used to identify musculoskeletal problems and direct potential solutions, similar to the Badger approach [Knapschaefer 1999].

Although ergonomics was initially included in the Badger BBS system to determine whether a hazard was present or not, the information gathered during observations was not sufficient to either identify specific risk factor exposures or control exposures not related to unsafe behaviors. For example, a person may use an awkward posture to do a task not because of an unsafe behavior but because the layout of the work station forces the worker to use an awkward posture. Typically, the observation of an unsafe behavior would result in training the worker not to use an awkward posture. However, because the awkward posture is a result of the work station layout and not a choice of method/behavior, further efforts are needed to resolve the risk factor exposures. In other words, observers require information for modifying tasks, equipment, tools, work stations, environments, and methods to eliminate exposures or use a hierarchical approach to control exposures (engineering controls, administrative controls, and PPE), with engineering controls being the preferred control measure [Chengalur et al. 2004]. Consequently, it was necessary to provide BBS observers with training not only in identifying specific risk factor exposures, but also in how to eliminate or control these exposures.

Training was provided to the BBS observers at both the Fairwater and Taylor Mines in July 2005 that focused on identifying risk factor exposures and presented simple ways to reduce or eliminate exposures associated with manual material handling. The training followed the observation process used by the observers to conduct safety observations and included role-playing exercises to allow the observers to be comfortable when doing ergonomic observations. To document risk factor exposures, an Ergonomic Observation Form was developed that also included simple ways to improve tasks. Information collected with this form includes risk factor exposures, body discomfort, root causes of the exposures, and corrective actions taken at the time of the observation. Practice completing the Ergonomic Observation Form was provided during the role-playing exercises.

In June 2006, additional training was provided to the BBS observers. This training consisted of a review of risk factors followed by additional practice at identifying risk factor exposures by viewing short videos and observing work tasks during field exercises. Methods to improve jobs were also discussed. Members of the Safety Teams also attended this training since these teams resolve observations not immediately addressed by the observers and CARE reports.

From August 2005 to May 2006, the BBS observers at both the Fairwater and Taylor Mines completed approximately 30 ergonomic observations. During 10 of the observations, the risk factor exposures were either resolved or job improvements were identified. The job improvements included PPE (antivibration gloves) and training on how to do a particular task without exposures to awkward postures, and engineering controls. Two examples of engineering controls included raising the work surface with saw horses, which allowed the use of neutral postures, and constructing a handtool to open covers on railcars, which eliminated bending the trunk and reduced the forceful exertion needed to release the latch.

Ergonomic observations are maintained in an electronic spreadsheet, which includes all of the fields on the observation forms and the status regarding action, if any, being taken to address the risk factor exposures. Additionally, interventions are being documented using a format to show how the task was done both before and after the intervention was implemented. Information on the intervention, such as cost and source (manufacturer), risk factor exposures, and body part affected are included in this document. The intervention forms are distributed to associates via hard copy and Intranet to encourage improvements in other jobs and to share information among Badger facilities. Posters highlighting interventions are also used to encourage associates to participate in the ergonomics process.

The process being implemented at Badger is proactive as it addresses exposures to risk factors and not just injuries. During the first year of this process, the emphasis has been on addressing CARE reports and BBS ergonomic observations. However, information learned by the associates during the Ergonomics and Risk Factor Awareness Training was also applied to the design of new work areas and facilities. Badger's process is participatory and as it matures will move to a more comprehensive process with the incorporation of ergonomic principles into more processes that affect employee safety and health.

> "Our ergonomics process has become a critical component of our overall safety program. Historically, ergonomic issues were the No. 1 cause of associate injury. Through this process, we are now able to proactively address ergonomic risk factors, resulting in a healthier, happier, more productive workforce. The process has also resulted in a significant reduction in lost time and reportable accidents."
>
> —*Marty Lehman, Safety Associate*
> *Badger Mining Corp.*

Vulcan Materials Co.

Vulcan Materials Co. is the largest U.S. producer of construction aggregates (crushed stone, sand and gravel). At yearend 2006, Vulcan had 372 facilities located in 21 states, the District of Columbia, and Mexico employing approximately 8,000 employees. The facilities are diverse in function, including stone quarries, sand and gravel plants, sales yards, asphalt plants, and ready-mix concrete plants. In 2006, Vulcan shipped 255.4 million tons of aggregates.

As a company, the basic organizations within Vulcan are seven autonomous divisions. The safety program is multilevel with Safety, Health and Environmental (SHE) Teams at the plant level, a Safety and Health Department at the division level (Safety Manager and Safety and Health (S&H) Representatives), and a Safety and Health Department at the corporate level (Safety Director and two safety professionals). Members of the plant SHE Teams include two to four hourly employees who volunteer for this assignment. The main functions of the SHE Teams are to conduct periodic inspections of the site and then report the findings to the Plant Manager. The division safety staff provide technical support to the plant management and SHE Teams, while the corporate safety staff provide technical support to the Division Safety Department.

In 2002, the National Stone, Sand and Gravel Association established a goal for its members to reduce their overall injury rate by 50% with 5 years. Vulcan committed to meeting this goal and immediately took steps to address safety and health hazards, which resulted in significant reductions in its injury rate. However, the injury rate was still above its goal because many of the injuries that were still occurring were a result of exposures to MSD risk factors. Vulcan decided it needed to take another approach. In August 2005, NIOSH researchers and Vulcan safety personnel (corporate- and division-level safety professionals) met to discuss how ergonomic principles could be applied within Vulcan Materials Co. to prevent MSDs. Because Vulcan has many facilities with fewer than 50 employees and limited on-site safety and health expertise, it was necessary to develop a plan to address both of these issues and also to address the overall size of the company. The plan that was developed took a two-phase approach. The first phase demonstrates how ergonomics can be applied at the Vulcan sites; the second phase lays the foundation for implementing a process throughout the company. To date, the first phase involved implementing ergonomics processes at two pilot sites within the Mideast Division. The second phase began with introducing ergonomic concepts and Vulcan's ergonomics initiative to other Vulcan sites.

At the pilot sites (North and Royal Stone Quarries), ergonomics was integrated with the existing safety and health programs, primarily with the company's "Taking Work out of Work" injury reduction initiative. Employees are encouraged to report risk factor exposures, using a risk factor report card, to the Ergonomics Review Team, whose members include the Plant Manager, the pit and plant supervisors, and the SHE Team leader. The Ergonomics Review Team, along with input from the S&H Representative, addresses the concerns using the process shown in Figure 4. When the concerns are investigated, a Manual Task Risk Assessment Form is used to evaluate risk factors, determine which risk factors should be controlled, and establish a prioritization score for determining which exposures should be addressed first.

The Vulcan process includes documenting the concern and the action taken to address the concern in a pilot database. As Vulcan expands its application of ergonomics throughout the Mideast Division and the other six divisions, information from the submitted cards and controls implemented will be captured in a division- or corporate-wide database and will be used as a resource for finding solutions to specific exposures, as well as to identify trends.

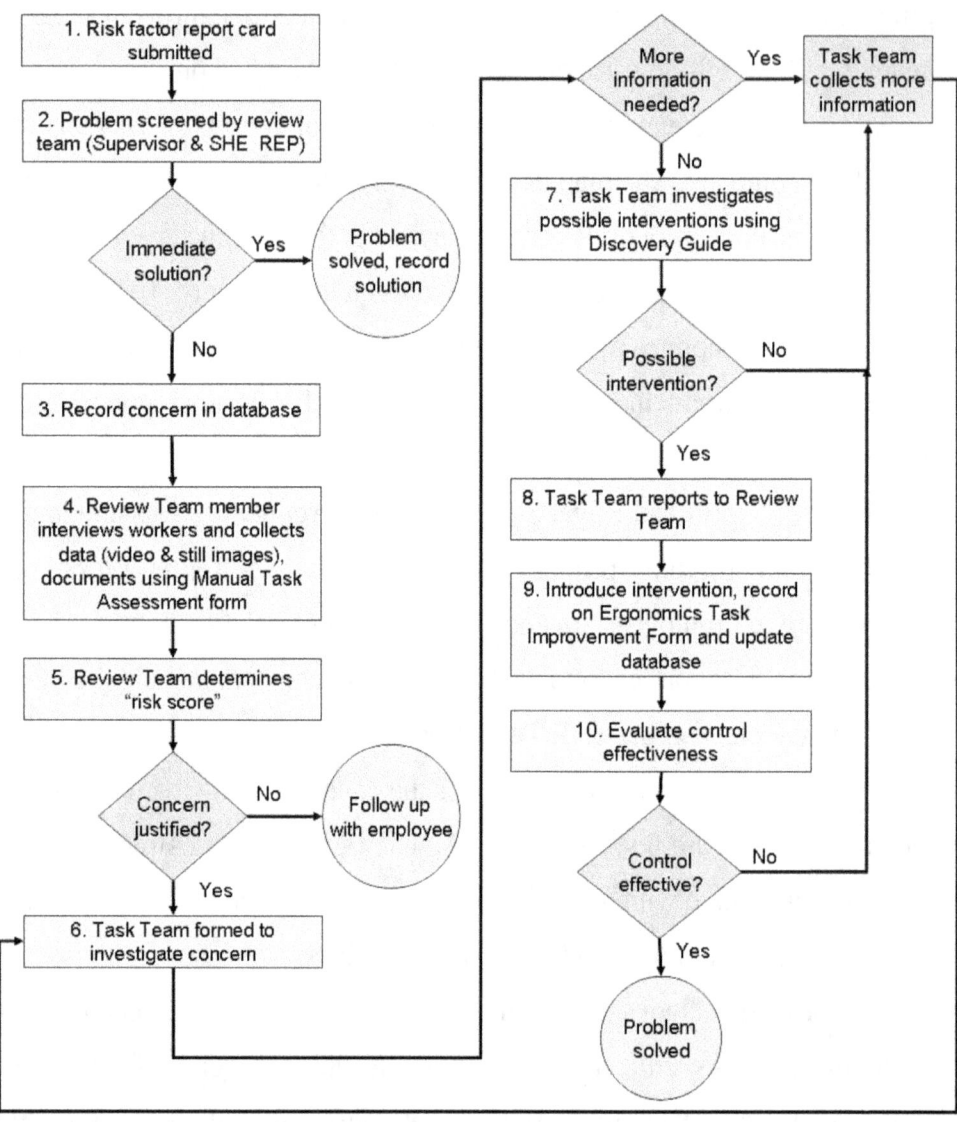

Figure 4.—Flow diagram of Vulcan Materials Co. task improvement process.

In April 2006, Vulcan employees at the two pilot sites received ergonomics and risk factor awareness training. The objectives of the employee training were twofold: to provide employees with skills for identifying risk factors in their work areas similar to their skills for identifying safety or health hazards, and to encourage employee participation in the ergonomics process. Prior studies have shown that an important element of successful ergonomics processes is employee involvement [Cohen et. al. 1997]. The employee training was given in two 90-minute sessions, 1 week apart, and was modified to include a homework assignment that encouraged employees to complete report cards identifying risk factor exposures for two tasks they do as part of their jobs. The first session of this training was given by the Division Safety Manager; the second session was given by the S&H Representatives assigned to the pilot sites. To become familiar with the training, these instructors attended a train-the-trainer session offered by NIOSH in February 2006.

The S&H Representatives assigned to the pilot sites and the Ergonomics Review Teams at both pilot sites were given additional training on implementing the ergonomics process, primarily how to process report cards, evaluate risk factor exposures, and determine appropriate controls. This training, given in April 2006, was a combination of classroom training and a field exercise. In the classroom, participants discussed how to implement the process, evaluate various implementation tools, and viewed several short videos to gain practice at identifying risk factor exposures. The field exercise provided practice with observing actual tasks being performed by employees and completing the Manual Task Risk Assessment Form. The field exercise was followed by a brainstorming session to determine solutions for the observed risk factor exposures.

In July 2006, the S&H Representatives and Ergonomics Review Team members were offered another training session focusing on job improvements, primarily selecting handtools and modifying manual tasks. Additional information was provided on the stress experienced by the back muscles and spinal discs during various lifting tasks. Participants were given practice at determining options for reducing exposures to risk factors by analyzing several tasks performed at their sites and then brainstorming job improvements.

Vulcan initiated the second phase of its application of ergonomic principles in November 2005 by offering all division S&H Representatives training that helped them to identify risk factor exposures and determine simple task improvements for reducing or eliminating risk factor exposures. During this training, the representatives were asked to submit examples of job improvements implemented at sites within their divisions. Approximately 10 improvements were submitted and posted on the Vulcan Intranet. In February 2006, NIOSH introduced ergonomic concepts to the Mideast Division Plant Managers. This presentation focused on Vulcan injury statistics with risk factor exposures and how ergonomics helped other companies to reduce their injury rates. The Mideast Division Engineering Department also received training from NIOSH in July 2006. This training emphasized the need to apply ergonomic principles during the planning and design stages to prevent exposures to risk factors. Specific components of this training included anthropometry and work station and conveyor design principles. For a homework assignment, participants were asked to design a sales yard clerk work station that could be used as a prototype for other Vulcan sites. The training/presentation offered during this phase was conducted primarily by NIOSH researchers, with support from Vulcan safety and health staff who provided information specific to Vulcan injury rates.

Because Vulcan is applying ergonomic principles at several levels within its company, there are several champions. At the pilot sites, the Plant Manager and the S&H Representatives are the champions. At the division level, the division Safety Manager is the champion. At the corporate level, the champion is the corporate Safety Manager.

> "At first, I had some concerns about the ergonomics process creating problems and I was resistant to the idea of implementing a process, even though our employees have been encouraged to improve their jobs with our 'Taking Work out of Work' Initiative. However, the ergonomics process has incorporated this initiative into a formal process and given our employees a green light to think out of the box to make their jobs easier."
>
> —Bryan Moore, Mideast Division Safety and Health Representative
> Vulcan Materials Co.

Lessons Learned

When implementing new processes, there are always lessons to be learned. Some of the lessons learned by the three companies in the above case studies included the following:

Bridger Coal Co.

Committee Participants: Early in the implementation phase, a number of leadership and committee members were replaced. The designated champion moved to a corporate position and a new champion had to be selected, and some committee members chosen to represent their departments either did not have the time or were not interested in being on the committee. Although some changes in membership are inevitable, it is important to select participants who want to be a part of the process and to allocate to them sufficient time to perform their work duties as well as committee responsibilities. This latter item was addressed by Bridger Coal Co. by including Ergonomics Committee participation in the job performance evaluations for salaried personnel and by altering employee schedules to permit sufficient time for committee activities. In addition, committee members supported each other by helping with tasks when other members did not have time to complete their assignments. While other companies have assigned a full-time coordinator to implement an ergonomics process, this was not considered necessary at the Jim Bridger Mine.

Process Development: There is no single "right" method that will work for all companies when developing a process. Although the Ergonomics Committee was given a lot of information and a number of ideas on how to proceed, it was necessary for committee members to determine what would work best to meet their needs. Because the committee had the responsibility for selecting the path it would take in implementing the process and ensuring its success, it was critical to have the right people on the committee, i.e., people who were interested in ergonomics, understood its value to the employees and to the company, and understood how to integrate it with other processes.

Process Implementation: Although employees received training after the Ergonomics Committee developed a procedure for submitting concerns, sufficient time was not allowed for the committee to become thoroughly familiar with the procedure. Then, because employee training resulted in the submission of numerous employee concerns, the committee was initially overwhelmed at the same time it was learning the procedure to address these concerns. Committee members were apprehensive about the amount of time needed to address all of the concerns and how the delay in responding would affect support for the process. Sufficient time should be given for a committee to become thoroughly familiar with its procedures prior to giving employee training and requesting that employees submit concerns.

Employee Training: When developing the employee training, several video clips were selected to demonstrate examples of risk factors. Approximately half of these video clips depicted Bridger Coal employees doing specific mining tasks. Unfortunately, other video clips taken at the Jim Bridger Mine did not adequately demonstrate risk factors, and clips from other operations were used. Some employees were critical that all of the video clips were not specific to work done at the Jim Bridger Mine. For future training, more video clips from the mine where employees are working, or from very similar mines, should be used.

Supervisory Training: Awareness training was focused mainly on employees and did not address the responsibilities of supervisors. Supervisors should receive additional training that specifically addresses their role in the ergonomics process. This training should demonstrate management's support for the process and should be done prior to the employee training so that the supervisor can express support for implementing the process. Supervisory training is particularly critical for supervisors who may have employees who are reluctant to participate. The concerns of these employees may never be addressed unless their supervisor initiates an action with the Ergonomics Committee. In addition, it is imperative that supervisors be fully aware of the way the company plans to conduct business related to ergonomic concerns.

Badger Mining Corp.

Associate Training: Following the training given to associates, it seemed that some associates still had a difficult time understanding the benefits of ergonomics and grasping the concept of how ergonomics can be used to make their jobs easier. To address this issue, a practical exercise was developed for the upcoming refresher training. The exercise requested that the attendees perform a simple task that had a risk factor exposure. Then, with the materials at hand, they had to find a way to modify the task that reduces the risk factor exposure and also results in an increase in productivity. This hands-on exercise provided these associates with an improved understanding of how to apply ergonomics. This same exercise continues to be used as an introduction to ergonomics whenever new employee training is given.

Ergonomics Observation Form: Because a BBS system was part of the safety and health program implemented at both Badger mines, it was decided to expand the scope of observations that identified ergonomics as an issue. (On the BBS Observation Form, an ergonomics issue was identified with a single checkmark next to the word "ergonomics". No other information was required; however, in some cases, the observers would include a brief description of the issue in the "comment" section of the form.) This expansion consisted of completing a one-page Ergonomics Observation Form that included additional information, such as risk factors, root causes, body parts with discomfort, and potential solutions to the exposures. All observers were given two training sessions, approximately 1 year apart, that focused on identifying risk factor exposures and then determining solutions for reducing the exposures. Simple job improvements were primarily for manual material-handling tasks. Practice with completing the Ergonomics Observation Form was also part of the training. Following both training sessions, the Ergonomics Observation Form was not widely used by the observers even though ergonomics issues were marked on the BBS Observation Form. The failure to use the Ergonomics Observation Form was related to a lack of time. Observers stated they did not even have sufficient time to meet their established goals for just the number of observations, without completing the Ergonomics Observation Form. The total number of observations completed for 2006 was about 50% of the stated goal. During this time period, both mines had record production levels without an increase in associates, putting further demands on available resources for implementation of the BBS program. Consequently, followup for ergonomics observations was generally completed by just a handful of observers. It is believed that just a few of the observers should have been selected to focus on ergonomic risk factor exposures and to follow up on observations identifying ergonomics issues. These observers could have received more in-depth training on identifying exposures, as well as methods to reduce or eliminate the exposures. Additionally, they could have served as a knowledgeable resource at each mine.

Process Champion: Because of the decentralized team structure at the mine sites, the progress made when implementing the ergonomics process was not always known or documented. Many of the interventions implemented were independent efforts of the individual teams and were not tracked by the ergonomics process defined in Figure 3. Consequently, information about these interventions had to be obtained in other ways. Additionally, because the champion assigned to the ergonomics process was physically located at the Resource Center and not at either of the mine sites, he was not always kept apprised of the efforts to implement various interventions. Even though the champion spent much of his time at both mine sites, it may have been helpful to assign an associate at each mine site to serve as a resource for the implementation process and to facilitate communication between the champion and the site teams, particularly for documentation purposes.

Vulcan Materials Co.

Documentation: It has been more difficult than expected to document the progress made at the pilot sites. Talking to the workers at the sites clearly revealed they have continued to apply ergonomics principles to their jobs and made them easier, but this information was not formally being captured. While documenting the interventions is not necessary for the process to be successful, a method for routinely documenting interventions would be useful so they can be shared with other sites. Possible solutions include: assigning this responsibility to a division-level S&H Representative, who would document interventions during periodic visits to the sites; or capturing this information during monthly safety meetings.

Recognition: To encourage the implementation of interventions, it is necessary to acknowledge workers who improved their jobs and to promote the value of such intervention efforts. This is now being done with a periodic one-page newsletter that highlights several implemented interventions that may be of interest to other sites.

Maintaining Interest: After risk factor exposures that were easy to fix were addressed, intervention efforts were directed to solutions that take longer to implement because of higher costs and more complex approval processes. Consequently, workers were not seeing anything being accomplished, and interest in the ergonomics process seemed to wane. Although ergonomics was still part of the culture, it was not being actively applied. To maintain interest, it is necessary to continue to promote an ergonomics process, similar to other safety and health programs.

Summary

Applying ergonomic principles within the mining industry has been shown to be a viable approach for addressing exposures to risk factors by implementing task improvements. All three companies who partnered with NIOSH to implement an ergonomics process were able to integrate ergonomics within their existing safety and health programs and to establish a systematic process to resolve ergonomic issues and implement task improvements. As it matures, the implementation process will move from addressing risk factor exposures and MSDs to incorporating ergonomic principles in the design of future work stations/equipment/tools and equipment specifications. Risk factor exposures will be proactively addressed in the design and planning stages, and ergonomics will automatically be an accepted way of doing business within the organization.

From the case studies presented, it is apparent that the ergonomics processes were successful because each implementation plan was modified to meet specific needs and address differences within each company, such as the demographic differences listed in Table 1 and cultural and organizational differences (how employee participation is encouraged and implemented, organizational structure, communication channels, etc.). Table 2 compares the three case studies with regard to how the implementation approach was modified. However, all three mining companies followed a basic framework or model that included the following critical elements:

- Assign a champion to promote and serve as an advocate and leader in applying ergonomic principles.
- Provide training to employees and organizational entities responsible for implementing the ergonomics process. The training should be customized to meet the roles played by each group in the implementation process.
- Develop a systematic process to identify and control risk factors associated with methods, tools, equipment, work stations, and environment. An example of a generic process for improving tasks is shown in Figure 5. If an exposure cannot be resolved, then it should be reviewed later as additional information or options become available.
- Track and document progress to demonstrate the benefits of the process, share interventions, and communicate lessons learned.
- Integrate ergonomics with other processes that affect worker safety and health, such as purchasing decisions, work schedules, modifications to existing facilities/equipment, and procedures. By doing this, costly reengineering efforts to correct problems with risk factor exposures can be avoided.

Table 2.—Comparison of the three case studies with regard to demographics, organizations, and implementation approaches

Factor	Company		
	Bridger	**Badger**	**Vulcan**
Implementation responsibility	Ergonomics Committee	Safety Team	SHE Team and division- and corporate-level support
Champion(s)	Safety Manager Ergonomics Coordinator	Safety Associate	Manager, Safety Services (corporate level) Manager, Safety & Health (Mideast Division) Plant Managers
Groups receiving training	Employees Ergonomics Committee	Employees BBS observers Safety Team	Pilot employees Pilot SHE Team S&H Representatives Mideast Division Plant Managers Mideast Division Safety Manager Mideast Division Engineering Department
Record-keeping	Spreadsheet for employee concerns	Spreadsheet for observations	Database for employee concerns and interventions
Communication	Posters and newsletter articles	Posters and flyers	Newsletter and Intranet

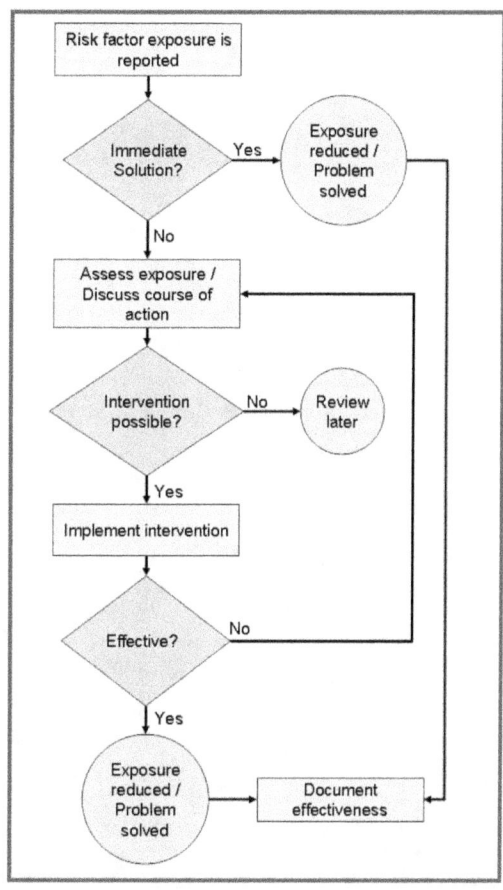

Figure 5.—Generic process for improving tasks.

The above framework is further described in a stand-alone two-page document in the Appendix. This document can be used to inform management as to what would be needed to implement an ergonomics process.

Section III
Process Effectiveness

Quantifying the effectiveness of ergonomics processes depends strongly on the organization and the original goal of the ergonomics process. It is common to see effectiveness measured in the number or incidence rate of workdays lost, number or incidence rates of injuries/illnesses, number of near-misses, or workers' compensation costs. Examples of how these measures have been used were reported by GAO [1997].

For some organizations, particularly small companies with limited injuries and illnesses, these measures may not be suitable. In these cases, use of survey tools, such as a Musculoskeletal Discomfort Survey form, may be more useful. Another constructive approach may be to quantify exposure levels to risk factors before and after implementing an intervention. For a lifting task, for example, the amount of weight lifted during a work shift may be measured before and after an intervention has been applied. Other examples include posture improvements, reducing the distance objects are carried, and reducing the number of repetitions performed. Other more technical tools that could be used to show reduced exposures include Rapid Upper Limb Assessment [McAtamney and Corlett 1993], Rapid Entire Body Assessment [Hignett and McAtamney 2000], the Revised NIOSH Lifting Equation [Waters et al. 1994], Hand Activity Level [ACGIH 2007b], and the Strain Index [Moore and Garg 1995]. Additional technical tools are available from Thomas E. Bernard, Ph.D., University of South Florida [Bernard 2007].

Since the three companies that partnered with NIOSH to implement ergonomics processes had very low injury incidence rates and few documented MSDs, it was not possible to use many of the above measures to demonstrate effectiveness. The companies were most interested in changing the way they looked at these types of injuries and tracking interventions, particularly when there were opportunities to share job improvements with other sites within the company. To assess process effectiveness for the three case studies, NIOSH used both discomfort data and/or interventions (job improvements) implemented. Discomfort data are presented for Bridger Coal Co., and summary information on interventions are presented for all three companies.

Bridger Coal Co.

Discomfort Survey

Reports of employee discomfort were obtained using a Musculoskeletal Discomfort Survey form adapted from the Standardized Nordic Questionnaire [Kuorinka et al. 1987]. The survey was administered in 2001 by NIOSH researchers and again in 2004 by Bridger Coal management. Although the survey was completed by 225 employees in 2001 and 116 in 2004, only 41 surveys could be matched for both years. The lower response rate in 2004 was attributed to a significant change in personnel (both turnover and reassignments) when the mine began converting its operations from surface to underground.

An analysis of the 41 matched survey reports did not indicate statistical differences in the rate of discomfort reported before and after the ergonomics process was implemented. However, the overall trend observed indicated a 17% decrease in discomfort reports following implementation of the ergonomics process. Fewer employees reported discomfort for the head, elbows, wrists/hands, upper back, and lower back (Figure 6). The most frequently reported body part with discomfort was the lower back both before and after the process implementation. Also, before the process was implemented more employees tended to experience discomfort in multiple body parts. For example, before the process was implemented 24% of the employees reported discomfort in three different body parts, while after the process was implemented only 17% reported such discomfort (Table 3).

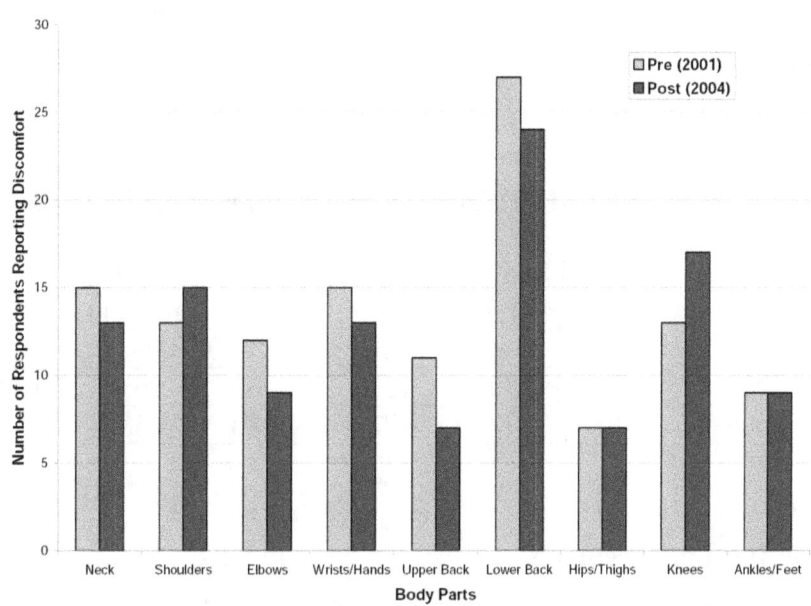

Figure 6.—Number of body part discomfort reports during the past 12 months for 2001 and 2004 (n = 41).

Table 3.—Number of Bridger employees reporting discomfort in one to four different body parts (n = 41)

	No. of body parts with discomfort				No. of employees reporting any discomfort
Year	1	2	3	4	
2001	7	5	10	8	37
2004	6	5	7	5	31

With regard to age, the percentage of employees reporting discomfort prior to the process was greater for three age groups (31–40, 41–50, and over 50) compared to the percentages for the same age groups after implementing the process. A slightly downward trend in reporting discomfort (83% to 72%) with increasing age (31–40 years old to over 50 years old) was observed after the process was implemented. A decrease in discomfort with increasing age was not observed before the process was implemented. The same results were observed when considering discomfort reports for the lower back (see Table 4).

Although trends in discomfort seemed to indicate that fewer employees were experiencing discomfort after the process was implemented, it is not possible to attribute the decline in discomfort directly to the ergonomics process. Many changes were occurring at the Jim Bridger Mine, such as job reassignments, that may have also impacted discomfort levels.

Table 4.—Percentage of Bridger employees reporting discomfort by age

Year	All body parts Age (years)			Back Age (years)		
	31–40	41–50	>50	31–40	41–50	>50
2001	100	89	100	86	65	100
2004	83	73	72	83	53	56

Interventions or Job Improvements

Employee reports of risk factor exposures and intervention efforts were documented and tracked by the Ergonomics Committee. Risk factor exposure data were obtained from employees who submitted concerns to the committee. Three years into the process the Ergonomics Committee received 55 concerns and successfully completed improvements for 22 concerns. Five more concerns were actively being addressed, and nine other concerns were on hold pending receipt of additional information. The remaining 19 concerns were either addressed as safety and health concerns or were not considered valid.

Table 5 provides information on interventions implemented by the Bridger Coal Co., including those initiated by the Ergonomics Committee. The average number of employees affected by an intervention was 16.8. Over half of the interventions involved the purchase of new equipment or seats. All of the purchases except one cost less than $3,000. Some modifications were completed by the equipment maintenance staff and did not result in significant expenditures of funds or time. The easiest type of concerns addressed by the committee involved rearranging equipment or work stations. Although many of the interventions seemed to be rather simple solutions, determining appropriate interventions usually involved detailed investigations and analyses to ensure employee acceptability and reduction in risk factor exposures. Activities performed when identifying and assessing potential interventions often included employee interviews, risk factor determinations, product identification and evaluation, and review of manufacturer approvals. Only one complaint could not be addressed by the committee because the intervention was considered cost-prohibitive.

Table 5.—Description and types of interventions completed by Bridger Coal Co.

Type of intervention	No. of employees affected	Brief description of intervention
Existing equipment modified	19	• Handle added to chocks to reduce back flexion
	5	• Loader foot pedal angle decreased to allow a relaxed foot position
	11	• Drill pedal moved to a more accessible location
	11	• Prill truck ladder handrail moved closer to the ladder to allow use with proper body positions
Work station rearranged	27	• Pump switch location changed to eliminate excessive reaching
	5	• Loader seat aligned with controls to eliminate twisting
New work stations purchased	2	• Adjustable office work stations purchased to allow proper body postures
New equipment purchased	14	• Lightweight welding helmets replaced heavier helmets to reduce the load supported by the neck and upper back
	14	• Wooden hammer handle with rubber guard replaced fiberglass handles to reduce hand vibrations
	9	• Nylon tie-down straps replaced heavier chains to reduce the load when handling the chains
	29	• Small table placed outside tool room so tools being returned could be placed on the table rather than on the floor. (Mechanics need to remove their safety glasses to use an eye scanner so they can gain access to the tool room. Holding tools hindered them from removing their safety glasses.)
	2	• Floor mats installed in warehouse to reduce discomfort from walking on concrete floors.
	16	• J-hook bar obtained to pull dragline cable rather than lifting the cable
	16	• Tractor purchased to move trailing dragline cable rather than moving the cable manually
	15	• Dragline work station improved with larger, more adjustable armrests and a footrest to reduce exposures to awkward postures
New seats purchased	47	• Seats changed in draglines, loaders, and blades to improve comfort
Availability of PPE improved	45	• Additional kneepads stocked in warehouse to reduce contact with hard surfaces

Badger Mining Corp.

Within 1 year of implementing its ergonomics process, Badger initiated more than 40 interventions, which are described in Tables 6 (Fairwater Mine) and 7 (Taylor Mine). Some of these interventions were planned prior to initiating the ergonomics process; however, information gained from the training led to improvements from the original design. All but a few of the improvements were engineering controls, and many of them involved obtaining new equipment or work stations. Some of the modifications to work stations or equipment were completed by the equipment maintenance staff and did not result in significant expenditures of funds or time.

Table 6.—Description and types of interventions completed by Badger Mining Corp.: Fairwater Mine

Type of intervention	No. of associates affected	Brief description of intervention
Existing equipment or work station modified	3	• Mirrors installed on mobile equipment to eliminate twisting when looking to the side and rear of the vehicle
	3	• Asphalt applied to unpaved roads to reduce whole-body vibration
	6	• Powered loading dock ramp replaced manual placement of dock ramp, eliminating forceful exertions
	6	• Automatic actuators installed in screen house replaced the requirement to manually reset actuators, which involved excessive reaching and back flexion
New work stations purchased or constructed	3	• Truck scale with washout system replaced manual cleanout while standing in a pit
	3	• Rail load-out canopy eliminated stooping under low-hanging equipment and improved protection from falls
New equipment purchased or constructed	3	• Brake stick used for railcars instead of climbing on railcar and manually setting brake, which involved forceful exertions
	3	• Floor mats purchased for dry plant to improve walking surfaces
	4	• Automatic greaser installed on vehicles replaced manual grease guns, eliminating awkward postures
	4	• Automatic grease gun replaced manual grease gun, which eliminated repetitive motions
	5	• Electric tarps replaced manual tarps on dump trucks, eliminating exposure to repetitive motions
	4	• Man lift replaced climbing ladders
	5	• Automatic dust collection screw replaced manually pounding on the hoppers
	3	• Tool to unlatch rail covers replaced manually unlatching the covers with a hand and foot, thereby avoiding excessive back flexion
New seats purchased	1	• New office chairs replaced existing chairs to promote improved postures
	2	• Air-ride seat installed in haul truck to improve postures and reduce whole-body vibration

Table 7.—Description and types of interventions completed by Badger Mining Corp.: Taylor Mine

Type of intervention	No. of associates affected	Brief description of intervention
Existing equipment or work station modified	6	• Rail cleanout facility modified to allow a standing posture rather than a stooped/squatting posture
	16	• Dozer operator compartment modified with an improved seat
	16	• Smaller 3.0-gallon buckets for preserving drilling samples replaced 5.0-gallon buckets, which reduced forceful exertions when removing buckets from holes
	16	• Ramp leading into pit widened
	16	• Haul roads straightened
	16	• Ride control installed on new loaders to reduce whole-body vibration
	7	• Airflow in dryer pipe revamped
Work station rearranged	5	• Tools placed in tool buckets so weight is evenly distributed and avoids leaning to one side
New work stations purchased or constructed	5	• Raised (waist-high) work station built for constructing bucket elevators to avoid working on floor and awkward postures
New equipment purchased	6	• Hy-vac truck purchased for rail cleanout replaced manual shoveling
	6	• 2-inch hose on Hy-vac replaced heavy 4-inch hose
	6	• Brake stick used for railcars instead of climbing on railcar and setting brake manually, which involved forceful exertions
	6	• Railcars with lightweight hatches replaced railcars with heavy metal covers, which reduced forceful exertions when lifting the covers
	7	• Autosamplers installed in dry house replaced manual collection of samples
	2	• Telephone headset purchased for receptionist to eliminate supporting the phone with the shoulders
	5	• Drills purchased for bucket elevator construction
	5	• Shock-absorbing hammers replaced regular hammers
	5	• Antifatigue mats placed in heavy traffic areas of the shop to reduce discomfort from walking on concrete floors
	1	• Wagons built to transport tools instead of carrying tools
	7	• Cable cutter attachment for drill replaced manual cutter, eliminating exposures to forceful exertions and repetitive motions

Table 7.—Description and types of interventions completed by Badger Mining Corp.: Taylor Mine—Continued

Type of intervention	No. of associates affected	Brief description of intervention
New equipment purchased— Continued	5	• New pickup trucks replaced Army surplus vehicles, which reduced whole-body vibration levels
	6	• Electric grease guns replaced manually operated grease guns, eliminating repetitive motions
	7	• Elevator installed in new dry plant, which replaced the need to climb stairs while carrying tools
	5	• Automatic parts washer replaced manual washing of parts, eliminating exposures to forceful exertions, repetitive motions, and stooped postures
	6	• Hinged screen covers replaced covers that had to be manually lifted off the screen housing, reducing forceful exertions
New seats purchased	16	• Replaced seat in drill to improve postures and reduce whole-body vibration levels
Elimination of equipment	6	• Railcars with trough hatches removed from service
Work practice modified	1	• Modified method to open bulk bags to eliminate stooping and leaning into bag
PPE	5	• Antivibration gloves purchased for constructing bucket elevators
	5	• Welding helmets with autodarkening lens replaced helmets with regular dark lens
	5	• Shoe insoles provided to maintenance workers to reduce discomfort when standing/walking on concrete floors

Vulcan Materials Co.

Immediately following the employee training, both Vulcan pilot sites implemented job improvements in response to the Risk Factor Report Cards submitted by the employees. Within 12 months, several interventions were completed at both pilot sites, as well as at Central Services, which received the Ergonomics and Risk Factor Awareness Training as Vulcan expanded the process within the Mideast Division. Although many of the interventions involved the purchase or construction of new equipment, few expenditures exceeded $5,000. In many cases, the labor was done internally and the costs of the interventions were insignificant.

Table 8.—Description and types of interventions completed by Vulcan Materials Co.

Type of intervention	No. of employees affected	Brief description of intervention
Existing equipment or work station modified	1	• Standing work station was converted to a sit-stand work station for the crusher operator
	4	• Screen storage racks were reoriented from vertical to horizontal storage so the screens could be placed on the rack with a forklift
	1	• Removable stairway added to drill to improve egress/ingress
	2	• Water slide added to conveyor that collects and removes spillage from floor of tower, which eliminated manually hosing area to remove spillage
	1	• Side-view mirrors placed on scraper to eliminate looking over the shoulder
	2	• Mirrors installed at supply bins to view back of trucks as the bins are filling, eliminating twisting the head and neck
Work stations rearranged	1	• Counters used to track number of loads dumped at the crusher were moved to eliminate excessive reaching
New work stations purchased or constructed	4	• Constructed ramps to replace manual jacking of vehicles when changing oil
	4	• Moved storage location for vehicle filters from second floor accessed via a stairway to a shed adjacent to the work area – eliminated climbing stairs and unsafe practice of holding bulky boxes when descending stairs
New equipment purchased or constructed	3	• Tool boxes placed on each level of the screen tower to eliminate carrying tools to different levels when repairs are needed
	3	• Water valves and hoses installed on all levels of screen towers to eliminate pulling the hoses to different levels of the towers
	3	• Crane installed to lift screens to the multiple levels of the screen towers rather than carrying the screens up several levels of stairs
	5	• Slide sledge was obtained to replace some uses of sledgehammers
	1	• Manual lifting and carrying of waste buckets were replaced with a waste bin modified so it could be moved with a forklift
	2	• Remote control to release materials from storage bins replaced manually pulling on cord

**Table 8.—Description and types of interventions completed by Vulcan Materials Co.
—Continued**

Type of intervention	No. of employees affected	Brief description of intervention
New equipment purchased or constructed— Continued	3	• Antifatigue mats purchased for crusher operator work station and workshops • Replaced manual torque wrenches with hydraulic torque wrenches
	4	• Obtained battery-operated screwdriver to replace manual screwdrivers
	4	• Obtained ¼-inch air wrench to replace manual screwdriver
	4	• Hilman rollers used to move differentials under loaders • Automatic washer replaced manual cleaning of parts • Constructed wheeled table with an elevated working surface to move parts to the automatic washer, which eliminated back flexion when picking up parts
	1	• Installed automatic belt sampler to collect stone samples, which eliminated manually carrying 5-gallon bucket from conveyor to pickup track
	1	• Installed blind-spot camera in the stockpile area, which eliminated twisting head and neck to view traffic
	4	• Obtained wagon to transport vehicle filters instead of carrying them • Obtained circular wheeled cart for moving 55-gallon drums • Obtained rotating engine stand to position parts being repaired
	1	• Purchased new rock breaker with improved ingress/egress and control options/locations
New seats purchased	1	• New seat purchased for the drill now that allows operator to place feet on the floor
	5	• New seats purchased for Komatsu haul trucks
Job enlargement	2	• Mechanic's job duties expanded to include operating vehicles (dozer, loader, and haul trucks)
PPE	2	• Shoe orthotics provided to mechanics to reduce discomfort from standing/walking on hard surfaces • Mechanics gloves provided for handling objects with sharp edges

Section IV
Implementation Tools

This section includes tools that were used by Bridger Coal, Badger Mining, and Vulcan Materials to implement their ergonomics processes. When applicable, information is also provided to describe the purpose of the tool, when to use the tool, and how to complete the tool. Because tools were modified throughout the course of implementing the three processes, only the latest version of the tool is provided. Electronic files for the tools are provided on the enclosed CD and can be modified for personal use as desired. Table 9 provides summary information about each tool, including how to administer the tool and the time required. Table 10 provides information on when to use the tools included in this section.

Table 9.—Summary information about each tool included in this section

	Tool	How to administer	Time required	Notes/Comments
A	Risk Factor Report Card	Self-administered by employees	5 minutes	This card is a very simple way for employees to report concerns they have about their jobs. Information from the card can be entered into a spreadsheet and tracked.
B	Musculoskeletal Discomfort Form	Self-administered questionnaire for employees; One-on-one interview; Group setting	5–10 minutes	The survey should be administered periodically to assess changes. For determining the effectiveness of a process, discomfort levels should be tracked yearly for at least 3–4 years. For specific task interventions, discomfort levels could be tracked sooner, perhaps at 6-month intervals.
C	General Risk Factor Exposure Checklist	Self-administered questionnaire for employees after instructions are read to the employees either individually or in a group	15–20 minutes	Can be used to obtain information from several employees about the same job or position. As part of reading the instructions, complete the first and second exposures listed as examples of how the checklist should be completed. The information collected can be used to prioritize jobs for improvements.
D	Ergonomics Observations Form	Completed by observer during observation/interview of worker being observed	10–20 minutes (depends on length of interview)	Can be used during observations conducted for a BBS process to collect additional information needed to determine appropriate job improvements. Can be used to track number of exposures observed and to prioritize task interventions.
E	Hand Tool Checklist	Completed by safety and health personnel	5–10 minutes	Compares handtools so each can be evaluated in terms of the ergonomic design features. All of the features are weighted equally.
F	Manual Task Risk Assessment Form	Completed by safety and health personnel when evaluating tasks	10–20 minutes	Provides a very basic risk assessment and ranking system for comparing risk before and after an intervention is implemented and for prioritizing interventions by body part affected.
G	Ergonomic Task	Completed by safety and health personnel or supervisor	10–20 minutes	When possible, include photographs of intervention or how the task was done both before and after the intervention.
H	Risk Factor Cards	—	—	Can be used as a handout during training to reinforce concepts taught during training or as a quick reference by safety and health personnel.
I	Sticker	—	—	Can be given to employees attending ergonomics training or used as an incentive to report exposures.

Table 10.—Brief description of when to use the tools included in this section

If you want to…		Tools
Obtain information about risk factor exposures from employees	A	Risk Factor Report Card
Prepare a baseline prior to implementing an intervention, including an ergonomics process for reducing MSD risk	B C	Musculoskeletal Discomfort Form General Risk Factor Exposure Checklist
Evaluate the effectiveness of an intervention	B C	Musculoskeletal Discomfort Form General Risk Factor Exposure Checklist
Obtain detailed information about risk factor exposures for a BBS process	D	Ergonomics Observations Form
Evaluate ergonomic features of handtools, powered and nonpowered	E	Hand Tool Checklist
Identify and assess risk factor exposures	F	Manual Task Risk Assessment Form
Assign risk level to risk factor exposures or job tasks	F	Manual Task Risk Assessment Form
Publicize the effectiveness of an intervention	G	Ergonomic Task Improvement Form
Promote application of material learned in training about risk factor exposures	H I	Risk Factor Cards Sticker

TOOL A
Risk Factor Report Card

Purpose

To encourage employee participation in the ergonomics process by providing a reporting mechanism for potential risk factor exposures and any body discomfort that may be related to the exposure.

When to Use It

The Risk Factor Report Card can be introduced to employees as a homework assignment during training. After training, employees can then use this tool to report their exposures and any body discomfort associated with risk factor exposures. The employees can also use it to provide input on how to change the task to reduce or eliminate the exposures. To promote reporting, the cards should be placed in areas where they are readily accessible to employees, such as lunch/break rooms, locker rooms, or posted on safety and health bulletin boards.

How to Use It

The information obtained from the Risk Factor Report Card can be used to target an intervention for the specific task identified on the card or to target interventions based on trends from information obtained from multiple cards. Examples of how information from multiple cards can be analyzed to target interventions are presented below for Risk Factor Report Cards submitted by Bridger Coal and Vulcan employees. To conduct these types of analyses, it would be useful to maintain the information obtained from the cards in a spreadsheet or database.

Bridger Coal Co.

The results of an analysis of concerns are shown in Figure 7. Of the 36 concerns processed by the Ergonomics Committee, one-third of the concerns were submitted by mechanics and another third by heavy equipment operators. The most frequently reported risk factor exposure was repetition, followed by heavy lifting and forceful gripping. The least reported exposure was vibration from using handtools. Discomfort was most frequently reported in the lower back and wrists/hands. These results indicate that interventions should be targeted for tasks done by mechanics that may require lifting or gripping tools or by heavy equipment operators who operate controls and sit for prolonged periods.

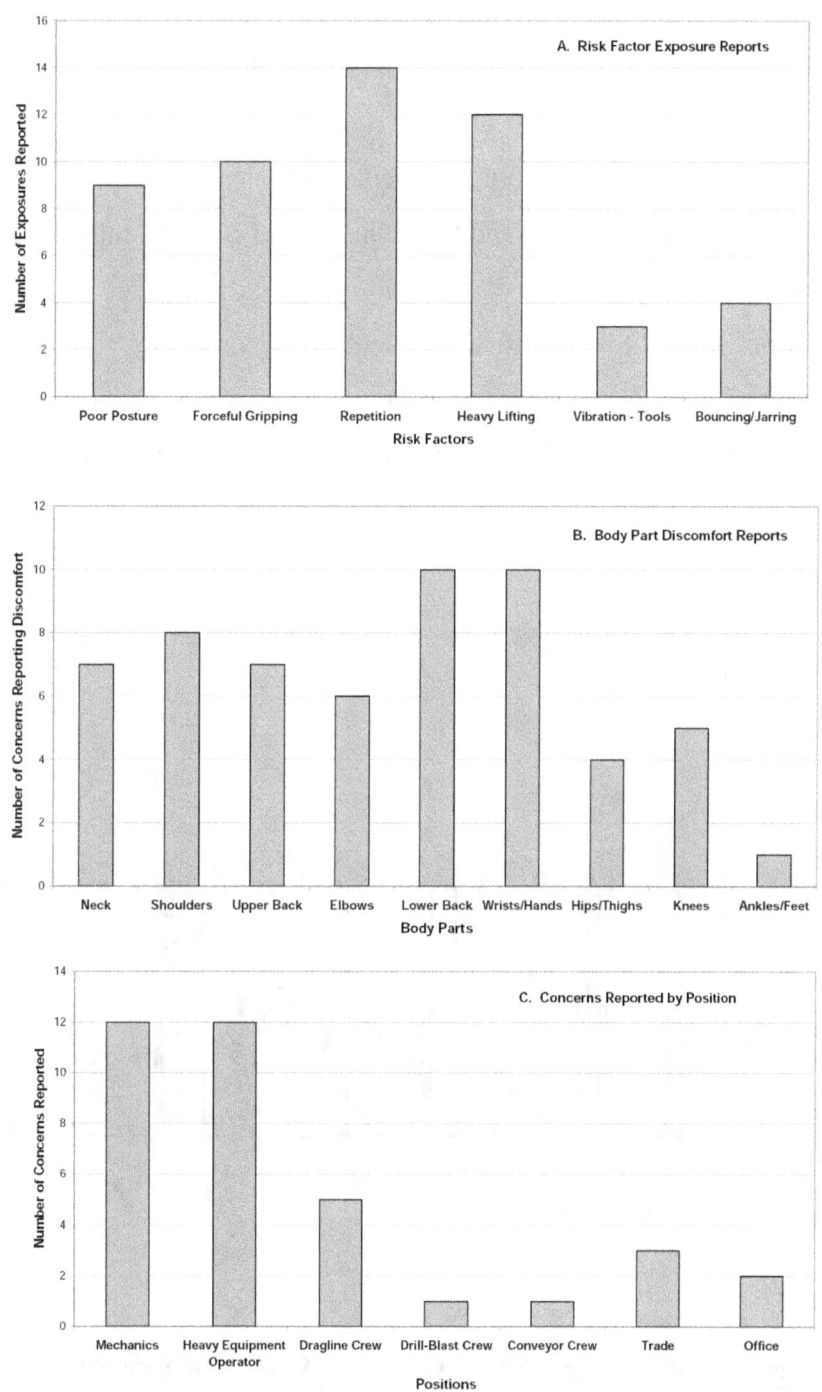

Figure 7.—Analysis of Bridger employee responses from submitted Risk Factor Report Cards.

Vulcan Materials Co.

As a homework assignment given during employee training, Vulcan employees submitted 42 report cards, 14 from the North Quarry and 28 from the Royal Stone Quarry. From the initial submittal of cards, risk factors and body discomfort were evaluated (Figures 8 and 9, respectively). At the North Quarry, poor postures, repetitive motions, and bouncing/jarring were the most frequently reported risk factors, while knees were the most frequently reported body part experiencing discomfort. By contrast, at Royal Stone, repetitive work and bouncing/jarring were the most frequently reported risk factors, while the lower back was the most frequently reported body part experiencing discomfort. Many of the reported exposures were associated with seating issues in heavy equipment. This information was used to initiate a study of whole-body vibration exposures from operating heavy equipment, primarily haul trucks and front-end loaders.

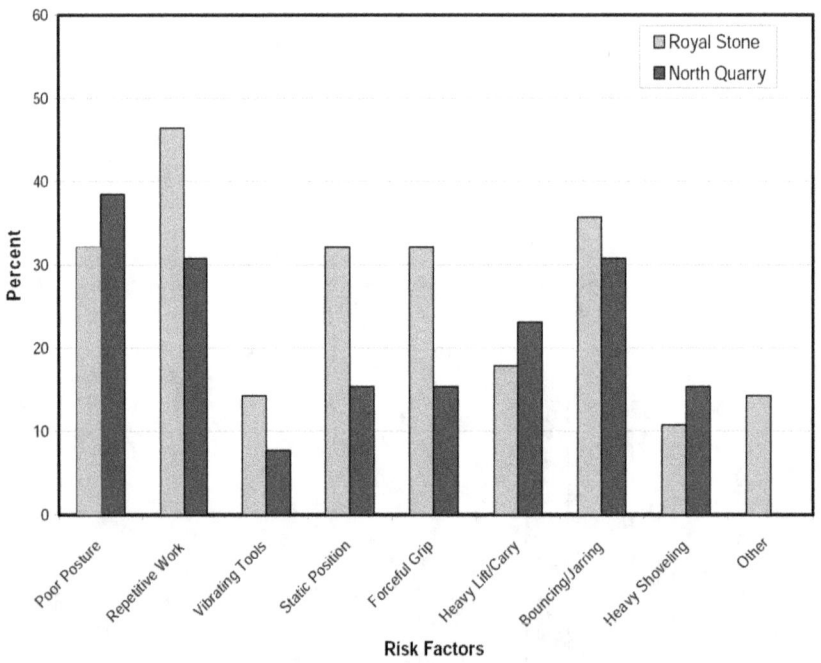

Figure 8.—Percentage of Risk Factor Report Cards identifying exposures to specific risk factors (more than one response permitted.)

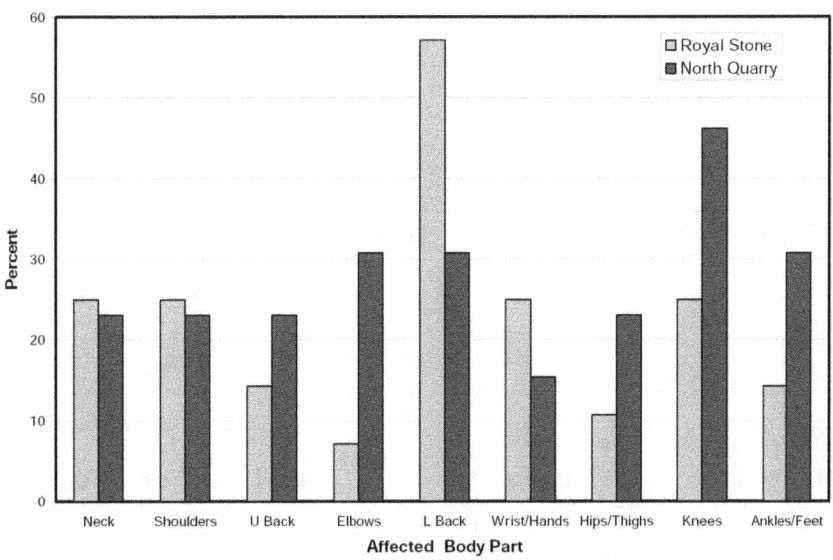

Figure 9.—Percentage of Risk Factor Report Cards identifying specific body parts with discomfort (more than one response permitted.)

How to Complete the Tool

1. Identify work area and/or job title.

2. Briefly describe the task. Provide enough information so that another person can understand the nature of the task.

3. Mark the risk factor exposures associated with the task. If the risk factor is not listed, identify or describe it on the "Other risk factors" line. (*NOTE:* The risk factors listed on the card are those common to mining tasks. The list can be modified if other risk factors are present.)

4. If discomfort is occurring, place an "X" on the body part in the diagram that is experiencing discomfort associated with the task.

5. In the "Comments/Suggestions" area, provide information that will be useful in evaluating the risk factor exposure. Examples may include: ways to improve the task, date when the discomfort started, and how is this task done differently than other similar tasks.

6. Indicate the name of the plant or mine. This line can be omitted if the ergonomics process involves only one site or mine.

RISK FACTOR REPORT CARD Name: _____

1. Work area: _____

2. Describe task: _____
   _____
   _____

3. Check all risk factors that apply:
   - ☐ Poor Posture   ☐ Forceful Gripping
   - ☐ Repetitive Work   ☐ Heavy Lifting/Carrying
   - ☐ Vibrating Tools   ☐ Bouncing/Jarring
   - ☐ Static Position   ☐ Heavy Shoveling

   Other risk factors: _____

4. Place X on affected areas.
   - Neck
   - Shoulders
   - Upper Back
   - Elbows
   - Lower Back
   - Wrist/Hands
   - Hips Thighs
   - Knees
   - Ankles/Feet

   Back View

5. Comments/suggestions: _____
   _____
   _____
   _____

6. Plant/Mine Name: _____

NOTE: The Risk Factor Report Card can be printed on 3 × 5 or 4 × 6 index cards.

TOOL B
Musculoskeletal Discomfort Form

Purpose

To identify the presence of discomfort by body part experienced by workers.

When to Use It

Use the Musculoskeletal Discomfort Form before and after implementing a process or a task specific intervention.

How to Use It

The Musculoskeletal Discomfort Form can be used to determine the effectiveness of an ergonomics process or a task specific intervention. The form is administered to employees to obtain a baseline prior to implementing a process or a task specific intervention, and then periodically after the process or task specific intervention has been implemented. The discomfort information can also be used to target interventions. For example, if several employees indicated they experienced shoulder pain, one could identify tasks that involve risk factor exposures for the shoulder, such as awkward postures or excessive force exertions, and then target those exposures for an intervention.

How to Complete the Tool

Employee ID: Indicate name or employee number.

Job/Position: Indicate job title or position.

How long have you been doing this job: Indicate number of years and months that you have worked in the job or position described above.

How many hours do you work each week: Indicate on average the number of hours worked per week.

Gender: Circle "M" for male, "F" for female.

Height: Indicate height in feet and inches.

Weight: Indicate weight in pounds.

To be answered by everyone (left column of table): For each body part listed, mark "No" if you have no discomfort or "Yes" if you have discomfort.

To be answered by those who have had trouble (discomfort) (middle and right columns of table):

Have you at any time during the last 12 months been prevented from doing your normal work because of the trouble? If you had discomfort any time during the past 12 months that prevented you from doing your normal work, mark "Yes" for that body part. If the discomfort did not prevent you from doing your normal work, mark "No."

Have you had trouble at any time during the last 7 days? If you had discomfort any time during the past 7 days, mark "Yes" for that body part. If the discomfort did not occur during the past 7 days, mark "No."

Musculoskeletal Discomfort Form (Based on the Nordic Questionnaire)

Employee ID: _____

Job/Position: _____ Gender: M F Age: _____ How long have you been doing this job? _____ years _____ months How many hours do you work each week? _____ Height: _____ ft. _____ in. Weight: _____

How to answer the questionnaire:

Picture: In this picture you can see the approximate position of the parts of the body referred to in the table. Limits are not sharply defined, and certain parts overlap. You should decide for yourself in which part you have or have had your trouble (if any).

Table: Please answer by putting an "X" in the appropriate box - one "X" for each question. You may be in doubt as to how to answer, but please do your best anyway. Note that column 1 of the questionnaire is to be answered even if you have never had trouble in any part of your body; columns 2 and 3 are to be answered if you answered yes in column 1.

To be answered by everyone	To be answered by those who have had trouble	
Have you at any time during **the last 12 months** had trouble (ache, pain, discomfort, numbness) in:	Have you at any time during the **last 12 months** been prevented from doing your normal work (at home or away from home) because of the trouble?	Have you had trouble at any time during **the last 7 days**?
Neck ☐ No ☐ Yes	☐ No ☐ Yes	☐ No ☐ Yes
Shoulders ☐ No ☐ Yes, right shoulder ☐ Yes, left shoulder ☐ Yes, both shoulders	☐ No ☐ Yes	☐ No ☐ Yes
Elbows ☐ No ☐ Yes, right elbow ☐ Yes, left elbow ☐ Yes, both elbows	☐ No ☐ Yes	☐ No ☐ Yes
Wrists/Hands ☐ No ☐ Yes, right wrist/hand ☐ Yes, left wrist/hand ☐ Yes, both wrists/hands	☐ No ☐ Yes	☐ No ☐ Yes
Upper Back ☐ No ☐ Yes	☐ No ☐ Yes	☐ No ☐ Yes
Lower Back (small of back) ☐ No ☐ Yes	☐ No ☐ Yes	☐ No ☐ Yes
One or Both Hips/Thighs ☐ No ☐ Yes	☐ No ☐ Yes	☐ No ☐ Yes
One or Both Knees ☐ No ☐ Yes	☐ No ☐ Yes	☐ No ☐ Yes
One or Both Ankles/Feet ☐ No ☐ Yes	☐ No ☐ Yes	☐ No ☐ Yes

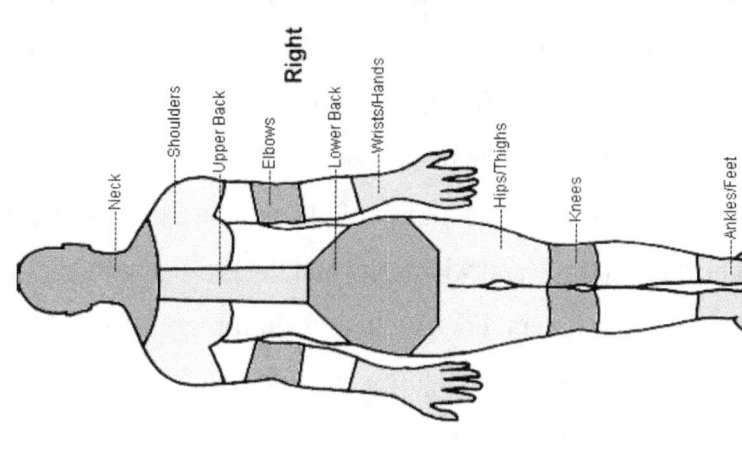

[Kuorinka et al. 1987]

TOOL C
General Risk Factor Exposure Checklist

Purpose

To obtain risk factor exposure profiles for specific jobs.

When to Use It

This tool is used prior to implementing a process or task-specific intervention to obtain a baseline for exposures. The checklist can then be administered periodically to determine if the exposures have been reduced or eliminated after the intervention has been implemented to determine the effectiveness of the process or intervention.

How to Use It

The information obtained will yield a listing of exposures associated with a job. This can be used to rank jobs based on exposures, providing a basis for prioritizing interventions. The checklist can also be used to categorize risk factor exposures by jobs, departments or teams, sites, or body part affected. This tool was designed to be completed by employees after receiving instructions from an administrator, such as a safety director. Once the exposures are known, then followup with employees is needed to determine the specific tasks associated with the risk factor exposures. The information obtained from the checklist can be used during brainstorming sessions with employees to facilitate discussion and focus the direction of the discussion.

An example of how this checklist was used at the Badger mines is shown in Tables 11 and 12. The percentage and number of employees reporting specific exposures were determined in order to identify the most frequently experienced risk factor exposures at each location (Table 11). The shaded cells indicate risk factor exposures that were reported by at least 60% of the employees at that site. This information could be used to determine which risk factors needed to be targeted at each site and which site had the greatest number of exposures affecting the greatest number of workers. For example, the information in Table 11 indicates that interventions may be needed to address the intensive keying and static postures for the Support employees and whole-body vibration exposures at Taylor, Fairwater, and Trucking. Risk factor exposures reported by employees on each work team at the two mines were also determined. The results for the Taylor Mine are shown in Table 12. These results identified the Mine Team as having the greatest number of employees with reported exposures (shoveling, whole-body vibration, and static postures). The Wash, Rail, and Maintenance Teams also had some risk factor exposures that were reported by 90% of the team members. This information allows one to target the intervention efforts to have the greatest impact on employee health after reducing or eliminating exposures.

A system was developed to obtain a single risk factor exposure score for each employee. The score is based on the number of exposures and severity of each exposure, according to duration of exposure, repetition of exposure, or weight lifted. All risk factors are scored equally, i.e., no risk factor is considered more important or weighted more than another risk factor.

Potential Issues With Checklist

Because of the large number of risk factors included in the checklist, the analysis can become daunting, particularly if a large number of checklists are completed. To simplify the analysis, it may be helpful to limit the risk factors to those affecting body parts of interest. For example, if back injuries are a problem, then only include the risk factors affecting the back. Using a software program (e.g., Microsoft Excel) to analyze the responses may also be helpful. For the Badger analysis, some of the risk factors were combined to simplify the analysis. In this case, the lifting risk factors were combined, as well as pushing and pulling risk factors.

Table 11.—Number and percentage of employees reporting risk factor exposures

(Shaded areas indicate risk factor exposures reported by at least 60% of employees at each location. N = total number of employees at each location; n = number of employees reporting exposure.)

Risk factor	Description	Badger location							
		Taylor Mine (N=50)		Fairwater Mine (N=25)		Trucking (N=15)		Support (N=44)	
		n	%	n	%	n	%	n	%
Forceful exertion	Lifting	31	62	14	56	7	47	9	20
	Shoveling	37	74	15	60	2	13	5	11
	Pinch grip	29	58	18	72	6	40	7	16
	Grasping	32	64	18	72	8	53	5	11
	Carrying	19	38	10	40	3	20	9	20
	Pushing	16	32	10	40	5	33	4	9
	Pulling	11	22	7	28	4	27	4	9
Awkward posture	Hands above head	23	46	13	52	5	33	4	9
	Neck bent	33	66	18	72	6	40	27	61
	Wrist bent	24	48	21	84	10	67	24	55
	Back bent (stooping)	30	60	17	68	11	73	4	9
	Squatting	22	44	13	52	7	47	4	9
	Kneeling	22	44	13	52	10	67	3	7
	Static position	26	52	16	64	11	73	40	91
Vibration	Moderate hand-arm	15	30	11	44	5	33	2	5
	High hand-arm	16	32	9	36	4	27	1	2
	Whole body	42	84	18	72	10	67	3	7
Contact stress		20	40	8	32	5	33	3	7
Intensive keying		18	36	11	44	2	13	40	91
Repetition		20	40	14	56	2	13	9	20

Table 12.—Number of employees reporting risk factor exposures by teams at the Taylor Mine

(Shaded areas indicate risk factor exposures reported by most employees for specific teams when the number of team members exceeds 1.
(N = total number of employees on team; n = number of associates reporting risk factor exposure)

Risk factor category	Specific risk factor	Taylor Teams								
		Dry N=7	Electrical N=1	Maintenance N=5	QTAT N=1	Rail N=6	Wash N=11	Mine N=16	Transload N=1	Operations N=2
		n	n	n	n	n	n	n	n	n
Forceful exertion	Lifting	2	1	5	0	6	8	7	1	1
	Shoveling	3	0	3	1	5	8	15	1	1
	Pinch grip	3	1	3	1	4	9	8	0	0
	Grasping	2	1	3	0	6	9	9	1	1
	Carrying	0	1	4	1	0	7	4	1	1
	Pushing	1	1	2	1	1	6	3	1	0
	Pulling	0	1	1	0	1	5	2	1	0
Awkward posture	Hands above head	0	1	3	0	3	8	6	1	1
	Neck bent	0	1	4	1	6	10	8	1	1
	Wrist bent	1	1	1	0	3	7	9	1	0
	Back bent (stooping)	0	1	4	1	6	10	7	1	0
	Squatting	0	1	2	0	6	7	6	0	0
	Kneeling	1	1	4	0	6	6	4	0	0
	Static position	1	1	1	1	0	6	13	1	2
Vibration	Moderate hand-arm	0	1	3	0	0	8	3	0	0
	High hand-arm	0	1	3	0	0	8	4	0	0
	Whole body	3	1	4	0	4	11	16	1	1
Contact stress		0	1	3	1	1	7	5	1	1
Intensive keying		2	1	0	0	2	8	2	1	2
Repetition		2	1	3	1	1	5	7	0	0

How to Complete the Tool

Because this tool is intended to be administered to employees, written instructions, shown below, have been prepared and are included with the tool.

Instructions

The purpose of completing this form is to identify exposures to MSD risk factors that occur when doing tasks required by your job. Your answers will be used to track the effectiveness of applying ergonomics at your mine.

1. **ID:** Name or employee number of employee completing checklist. (It is important to enter your name on the checklist so your checklist can be matched to future checklists.)
2. **Job/Position:** Provide the name of your job. Please be as specific as possible (mechanic, dozer operator, crusher operator, etc.).
3. **Team/Department:** Indicate the name of your organization.
4. **Date:** Date checklist is completed.
5. **Mine/Plant:** Name of your mine or plant.
6. **Shift:** Check the length of your typical shift.
7. **Brief Description of Your Job:** Provide a list of the main tasks you do for your job. Also list the number of hours/shift you spend doing each task and any equipment or tools you use. For example:

Main Tasks	Number of Hours	Tools/Equipment Used
Operate dozer	4 hours	CAT D10
Repair truck brakes	3 hours	pneumatic wrench

Risk Factors: Read each description of the risk factors while thinking about all the tasks you do that are a part of your job. Mark the choice that best applies to your job with a ✓ *or* **X**. If you do not perform the risk factor described, mark "Never." If you do the risk factor described periodically (once/week or once/month), then mark "Occasionally."

Some risk factors ask for additional information. Please write your response in the space provided.

On the last page of the checklist, list two tasks that you do for your job that you believe are the most physically demanding. Physically demanding means a lot of effort is required to do the task, or it involves one or more of the risk factors listed in this checklist.

General Risk Factor Exposure Checklist   ID _____

Job / Position _____ Date _____
Team / Department _____ Mine/Plant _____
Shift _____8 hrs   _____10 hrs   _____12 hrs   _____Other (describe) _____

Brief Description of Your Job

Main Tasks	Number of Hours	Tools/Equipment Used

Heavy or Frequent Lifting / Lowering / Shoveling

	Lifting *or* lowering object weighing more than 75 pounds: ___ Never ___Occasionally ___ less than one time per day ___ one or more times per day Lifting *or* lowering object weighing 55 to 75 pounds: ___ Never ___Occasionally ___ less than 10 times per day ___ more than 10 times per day	**Back / Shoulders**
	Lifting *or* lowering object weighing more than 25 pounds: ___ Never ___Occasionally ___ less than 25 times per day ___ more than 25 times per day	**Back / Shoulders**
	Lifting *or* lowering object weighing more than 10 pounds: ___ Never ___Occasionally ____less than 2 hours total per day ____more than 2 hours total per day	**Back / Shoulders**
	Shoveling: ____ Never ___Occasionally ____less than 1 hour total per day ____from 1 to 2 hours total per day ____more than 2 hours total per day What material do you shovel?_____	**Back / Shoulders / Arms**

Awkward Postures

	Working with the hand(s) above the head *or* the elbow(s) above the shoulders: ____ Never ____ Occasionally ____ for less than 2 hours per day ____ from 2 to 4 hours total per day ____ more than 4 hours total per day	**Shoulders**
	Working with the neck bent more than 30 degrees (without support): ____ Never ____ Occasionally ____ for less than 2 hours per day ____ from 2 to 4 hours total per day ____ more than 4 hours total per day	**Neck**
	Working with a bent wrist(s) – flexion, extension or deviation: Indicate Posture ____ Never ____ Occasionally _____ ____ for less than 2 hours per day _____ ____ from 2 to 4 hours total per day _____ ____ more than 4 hours total per day _____	**Wrists / Arms**
	Working with the back bent more than 30 degrees (without support): ____ Never ____ Occasionally ____ for less than 2 hours per day ____ from 2 to 4 hours total per day ____ more than 4 hours total per day	**Back**
	Squatting: ____ Never ____ Occasionally ____ for less than 2 hours per day ____ from 2 to 4 hours total per day ____ more than 4 hours total per day Kneeling: ____ Never ____ Occasionally ____ for less than 2 hours per day ____ from 2 to 4 hours total per day ____ more than 4 hours total per day	**Knees**

High Hand Force - Pinch Grip

Pinching an unsupported object:

____ Never   ____ Occasionally
____ for less than 2 hours per day
____ from 2 to 4 hours total per day
____ more than 4 hours total per day

What object do you pick up with a pinch grip?
_____

Elbows / Wrists / Hands

Pinch grip + wrists bent (flexion, extension, *or* in deviation):

____ Never   ____ Occasionally
____ for less than 2 hours per day
____ from 2 to 4 hours total per day
____ more than 4 hours total per day

Elbows / Wrists / Hands

High Hand Force - Grasp or Power Grip

Grasping an unsupported object(s) weighing 10 or more pounds per hand, *or* grasping with a forceful grip:

____ Never   ____ Occasionally
____ for less than 2 hours per day
____ from 2 to 4 hours total per day
____ more than 4 hours total per day

Elbows / Wrists / Hands

Grasping *plus* wrists bent (flexion, extension, *or* in deviation):

____ Never   ____ Occasionally
____ for less than 2 hours per day
____ from 2 to 4 hours total per day
____ more than 4 hours total per day

Elbows / Wrists / Hands

Highly Repetitive Work

Repeating the same motion (excluding keying activities) with little or no variation every few seconds:

    ____ Never   ___Occasionally
    ____ for less than 2 hours per day
    ____ from 2 to 6 hours total per day
    ____ more than 6 hours total per day

Shoulders / Wrists / Arms

Repeating the same motion (excluding keying activities) with little or no variation every few seconds *plus* wrists bent (flexion, extension, *or* in deviation) *plus* high, forceful exertions with the hands:

    ____ Never   ___Occasionally
    ____ for less than 2 hours per day
    ____ from 2 to 4 hours total per day
    ____ more than 4 hours total per day

Performing intensive keying (perform only keying with few or no breaks):

    ____ Never   ___Occasionally
    ____ for less than 4 hours per day
    ____ from 4 to 7 hours total per day
    ____ more than 7 hours total per day

Arms / Wrists / Shoulders / Neck

Performing intensive keying *plus* wrists bent (flexion, extension, *or* in deviation):

    ____ Never   ___Occasionally
    ____ for less than 2 hours per day
    ____ from 2 to 4 hours total per day
    ____ more than 4 hours total per day

Vibrating Tools (Hand-Arm Vibration)

Using grinders, sanders, jigsaws, or other handtools that typically have moderate vibration levels:

    ____ Never   ___Occasionally
    ____ for less than 2 hours per day
    ____ from 2 to 4 hours total per day
    ____ more than 4 hours total per day

Arms / Wrists / Shoulders

Using impact wrenches, chain saws, percussive tools (jackhammers, scalers, chipping hammers), or other tools that typically have high vibration levels:

    ____ Never   ___Occasionally
    ____ for less than 30 minutes total per day
    ____ for more than 30 minutes total per day

Arms / Wrists / Shoulders / Back

Bouncing or Jarring (Whole-Body Vibration)

Operating mobile equipment:
- ____ Never   ____ Occasionally   List equipment:
- ____ for less than 2 hours per day   _____
- ____ from 2 to 4 hours total per day   _____
- ____ more than 4 hours total per day   _____

I travel over rough roads (circle one):
Never   Sometimes   Most of the time   All of the time

Back / Hips / Legs

Contact or Impact Stress

Contacting hard or sharp objects like work surface edges or narrow tool handles, or striking an object with a hammer:
- ____ Never   ____ Occasionally
- ____ for less than 2 hours per day
- ____ from 2 to 4 hours total per day
- ____ more than 4 hours total per day

Describe sharp object / hammer _____

Shoulders / Elbows / Wrists / Arms

Static Postures

Standing without changing posture:
- ____ Never   ____ Occasionally
- ____ for less than 2 hours per day
- ____ from 2 to 4 hours total per day
- ____ more than 4 hours total per day

Sitting without changing posture:
- ____ Never   ____ Occasionally
- ____ for less than 2 hours per day
- ____ from 2 to 4 hours total per day
- ____ more than 4 hours total per day

Back / Hips / Legs

Carrying

Carrying objects *more* than 7 feet - check weight *and* frequency for most difficult carry (check "Never" if you do not carry objects):

OBJECT WEIGHT
- ___ Less than 20 pounds
- ___ 21 to 35 pounds
- ___ 36 to 50 pounds
- ___ More than 50 pounds

FREQUENCY
- ___ Occasionally   ___ Never
- ___ Less than 1 carry/minute
- ___ 1–2 carries/minute
- ___ 3–6 carries/minute
- ___ More than 6 carries/minute

Back / Shoulders / Elbows / Legs

Pushing and Pulling

Pushing against an object, such as a cart or handle, with a maximum effort (body leaning with bent legs into the push):
　　____ Never 　　____ Occasionally
　　____ less than 8 times per day
　　____ from 8 to 30 times per day
　　____ more than 30 times per day

Pushing against an object, such as a cart or handle, with a moderate effort (body slightly leaning with straight legs into the push, similar to pushing a full grocery cart):
　　____ Never 　　____ Occasionally
　　____ less than 16 times per day
　　____ from 16 to 50 times per day
　　____ more than 50 times per day

Back / Shoulders / Elbows / Legs

Pulling against an object, such as an electrical cable, fuel hose, cart, or handle, with a maximum effort (body leaning with bent legs into the pull):
　　____ Never 　　____ Occasionally
　　____ less than 8 times per day
　　____ from 8 to 30 times per day
　　____ more than 30 times per day

Pulling against an object, such as an electrical cable, fuel hose, cart, or handle, with a moderate effort (body slightly leaning with straight legs into the pull, similar to pulling a full grocery cart):
　　____ Never 　　____ Occasionally
　　____ less than 16 times per day
　　____ from 16 to 50 times per day
　　____ more than 50 times per day

Back / Shoulders / Elbows / Legs

Most Difficult or Physically Demanding Tasks
(Please provide a brief description of each task)

Why is this task difficult?

1.

2.

How to Score the Risk Factor Exposures

RISK FACTORS	SCORE
o Lifting or lowering object weighing more than 75 pounds o Lifting or lowering object weighing more than 55 pounds o Lifting or lowering object weighing more than 25 pounds	0 never 1 occasionally 2 less than 1 time per day 3 more than 1 time per day
o Lifting or lowering object weighing more than 10 pounds	0 never 1 occasionally 2 less than 2 hours total per day 3 more than 2 hours total per day
o Shoveling more than 5 pounds if done more than 3 times per minute	0 never 1 occasionally 2 less than 1 hour total per day 3 from 1 to 2 hours total per day 4 more than 2 hours total per day
o Working with the hand(s) above the head or the elbow(s) above the shoulders o Working with the neck bent more than 30 degrees o Working with bent wrist(s) o Working with the back bent o Squatting o Kneeling o Pinching an unsupported object o Pinch grip + wrists bent (flexion, extension, *or* deviation) o Grasping unsupported object(s) weighing 10 or more pounds per hand *or* grasping with a forceful grip o Grasping *plus* wrists bent (flexion, extension, *or* deviation) o Repeating the same motion (excluding keying activities) with little or no variation every few seconds *plus* wrists bent (flexion, extension, *or* deviation) *plus* high, forceful exertions with the hands o Performing intensive keying *plus* wrists bent (flexion, extension, *or* deviation) o Using grinders, sanders, jigsaws, or other handtools that typically have moderate vibration levels o Operating mobile equipment o Contacting hard or sharp objects like work surface edges or narrow tool handles o Sitting without changing posture o Standing without changing posture	0 never 1 occasionally 2 for less than 2 hours per day 3 from 2 to 4 hours total per day 4 more than 4 hours total per day
o Repeating the same motion (excluding keying activities) with little or no variation every few seconds	0 never 1 occasionally 2 for less than 2 hours per day 3 from 2 to 6 hours total per day 4 More than 6 hours total per day
o Perform intensive keying	0 never 1 occasionally 2 for less than 4 hours per day 3 from 4 to 7 hours total per day 4 more than 7 hours total per day
o Using impact wrenches, chain saws, percussive tools (jackhammers, scalers, chipping hammers), or other tools that typically have high vibration levels	0 never 1 occasionally 2 for less than 30 minutes total per day 3 for more than 30 minutes total per day

o	Pushing against an object, such as a cart or handle, with a maximum effort (body leaning with bent legs into the push)	0	if not checked
		1	less than 8 times per day
o	Pulling against an object, such as a cart or handle, with a maximum effort (body leaning with bent legs into the pull)	2	from 8 to 30 times per day
		3	more than 30 times per day
o	Pushing against an object, such as a cart or handle, with a moderate effort (body slightly leaning with straight legs into the push, similar to pushing a full grocery cart)	0	never
		1	occasionally
		2	less than 16 times per day
o	Pulling against an object, such as a cart or handle, with a moderate effort (body slightly leaning with straight legs into the pull, similar to pulling a full grocery cart)	3	from 16 to 50 times per day
		4	more than 50 times per day
o	Carrying objects more than 7 feet	0	never
		1	less than 35 pounds
		2	21 to 35 pounds
		3	36 to 50 pounds
		4	more than 50 pounds
		1	occasionally
		2	less than 1 carry/minute
		3	1–2 carries/minute
		4	3–6 carries/minute
		5	more than 6 carries/minute

Total score for each employee = Sum of scores for each risk factor

NOTE: This checklist was based on the Washington State Caution and Hazard Zone Checklists and the Followup Physical Risk Factor Checklist, but modified to be more applicable to mining and to be completed by employees. This checklist has not been statistically validated and should only be used as a guide.

TOOL D
Ergonomics Observations

Purpose

To identify risk factor exposures and subsequent actions to reduce or eliminate exposures.

When to Use It

This tool can be used to collect exposure information when observing work tasks as part of a risk assessment. It can also be used when observing tasks as part of a BBS process. Because many risk factor exposures result from inadequately designed equipment, tools, and work stations and not from an unsafe behavior, it is important to capture information that will allow action to be taken to correct the root cause of the exposure.

How to Use It

This form can be used to track information about risk factor exposures. It can be used specifically to track the type of risk factor exposures, the occurrence of body discomfort, and the root cause of the exposure. This form can also be used to document simple improvements taken to reduce or eliminate the exposures. Because the form also asks the observer to rate the level of risk he or she believes is associated with the risk factor exposure, it can be used as a very basic prioritization method. For example, exposures rated with "very high risk" would have a higher priority for an intervention than those rated with "low risk."

How to Complete the Tool

Mine: List name of mine.

Location: List geographic location of mine (name of nearest town/city).

Team/Department: List name of team or department that is the subject of the observation.

Task: Briefly describe the task being observed.

Time: Indicate the time of the observation.

Date: Indicate the date of the observation.

 # Observed: Indicate the number of employees being observed.

Observer: Indicate the name of the person doing the observation.

1. Indicate the risk factor exposure(s) observed by marking the box next to the risk factor. If the risk factor is not listed, check the "Other" box and briefly describe the risk factor exposure.

2. Indicate how often the task is performed to determine the frequency of the exposure. Determine if the task is done on a regular basis or seasonally. For tasks that are done on a regular basis:
   - Indicate the number of times a task is done per shift. For example, if the task is done six times during a shift, complete this question as "6 times/shift."
   - Indicate how many shifts this task is performed per week, month, *or* year.
     - If the task is done every week, indicate the number of shifts the task is performed per week.
     - If the task is not done every week, but is done every month, then indicate the number of shifts per month the task is performed. For example, if the task is done seven shifts per month, then complete this question as "7 shifts/month."
     - If the task is not done every month, then indicate the number of shifts per year the task is performed. For example, if the task is done 15 shifts per year, then complete this question as "15 shifts/year."

   For tasks that are performed on a seasonal basis:
   - Check the box to indicate that the task is seasonal, and then record the number of weeks per year the task is performed.
   - Indicate the number of times a task is done per shift. For example, if the task is done 12 times during a shift, complete this question as "12 times/shift."
   - Indicate the number of shifts the task is performed per week. For example, if the task is done three shifts during a week, complete this question as "3 shifts/week."

3. Indicate if the employee being observed is experiencing any discomfort, and then mark the body part with discomfort by placing an "X" on the body part in the diagram. This information will not only identify the development of potential MSDs, but can help focus job improvements.

4. Check the root cause of the risk factor (why the risk factor is occurring), and give a brief description of the root cause. Identifying the root cause can provide direction on how to change the task to reduce the risk factor exposures. For example:
    - If the employee is required to lift an object weighing 90 pounds, then check "effort or strength required" as the root cause, and then briefly describe it – "lifted object weighs 90 pounds."
    - If the task is repeated throughout the shift, such as operating a loader, then check "cycle time" as the root cause, and then describe it – "time to fill and dump one bucket is 20 seconds"; and also check "duration of task" as another root cause, and then describe it – "loader is operated for 7 hours per shift."
5. Rate the degree of risk of the exposure as either "none," "low," "medium," "high," or "very high." This rating is based on understanding the level of exposure, how often the exposure occurs, and the duration of the exposure. For example, if the exposure is from lifting an object weighing 30 pounds once a week, then the risk would be rated as "low." If the exposure is from lifting an object weighing 100 pounds once a week, then the risk would be rated as "high." The conditions provided in Section I under "Basic Elements of Ergonomics Risk Management Processes" can be used as a guide for ranking the task. If any of the conditions are present, the task would be rated as either "medium," "high," or "very high" risk. (This risk ranking is only meant as a very crude attempt to set a prioritization when several exposures may need to be addressed and limited resources are available to address the exposures.)
6. Indicate if there is another way to do the task that would reduce or eliminate the exposure and describe how the task can be changed. This question provides the observer with an opportunity to obtain ideas from the worker about ways to change the task.
7. Indicate if the exposure was reduced or eliminated at the time of the observation.
8. Provide any pertinent comments related to the exposure, discomfort, or ways to control the exposure.
9. Indicate any of the suggested job improvements that were discussed or tried at the time of the observation. This information will be useful for determining appropriate followup to resolve the exposure.

ERGONOMICS OBSERVATIONS

Mine: _____ Location: _____ Team: _____ Task: _____

Time: _____ Date: _____ # Observed: _____ Observer: _____

1. Check all the risk factors that you observe.
   - ☐ forceful exertion ☐ repetition ☐ static posture ☐ awkward postures ☐ contact stress
   - ☐ pressure points ☐ vibration ☐ other_____

2. How often is this task performed? _____times/shift _____shifts/week
   ☐ Seasonal ___weeks/year _____shifts/month _____shifts/year

3. Does employee experience any discomfort when doing this task? Yes   No
   If yes, what body parts have discomfort? (Place X on body part)

4. What is causing the risk factors? Check (√) root cause (s) and give brief description.

CHECK	ROOT CAUSE	BRIEF DESCRIPTION
	Effort or strength required	
	Location of parts, equipment or tools	
	Position of parts, equipment or tools (How the part is positioned in reference to the worker)	
	Design of parts, equipment or tools	
	Frequency of task (how often is task done)	
	Duration of task (how long it takes to do task)	
	Productivity levels	
	Method used / required to do the task	
	Training not adequate / need training	
	PPE not available / wrong PPE used	
	Environment – heat, cold, restricted space, etc.	
	Other	

5. Rate the degree of risk factor exposures for the observed task: None   Low   Medium   High   Very High

COMPLETE ONLY IF EXPOSURES ARE RATED MEDIUM, HIGH or VERY HIGH

6. Is there another way to do this task that eliminates/reduces the risk factor(s)?   Yes   No
   If yes, describe how the task can be changed: _____

7. Was the risk factor exposure resolved at the time of the observation?   Yes   No

8. Comments: _____

9. Indicate options discussed by checking the box next to the solutions shown below and on back side of page.

Personal Protective Equipment	Administrative Controls
☐ Anti-vibration gloves – reduce vibration transmission ☐ Knee pads – reduce pressure points ☐ Shoe inserts – reduce foot discomfort and fatigue ☐ Cooling devices – reduce body temperature increases ☐ Cold weather clothing	☐ Job enlargement ☐ Job rotation ☐ Work pace and duration ☐ Work-rest cycles ☐ Training ☐ Shift schedule / overtime ☐ Exercise / stretches

MMH (Lifting) – Think Smart	MMH (Push/Pull) – Think Smart	MMH Design Strategies
☐ Plan activity ☐ Keep loads clos body ☐ Use lifting as ☐ Ask for assista *Balance load/Stronger person on bottom/Talk* ☐ One-handed lift / carry *Avoid/Use other hand as counterbalance/Alternate Hands* ☐ Avoid bending at the waist to lift objects ☐ Avoid twisting – take a step and turn	☐ Use push instead of pull ☐ Keep elbows near 90° ☐ Provide clear path ☐ Avoid slopes ☐ Avoid uneven floors	☐ Avoid manual handling ☐ Use mechanical aids (hand trucks/carts) ☐ Use carts with large casters ☐ Modify workplace *Store heavy items between knees & shoulders – Avoid placing on floor* *Store light items on top shelf* ☐ Decrease object/container size ☐ Decrease object/container weight ☐ Change container shape ☐ Add handles
Reducing Forceful Exertions Use ☐ Power tools ☐ Fixtures ☐ Slides & rollers ☐ Mechanical aids ☐ Gravity to move materials ☐ Leverage ☐ Power grip – not pinch grip	**Reducing Excessive Motions** ☐ Use power tools ☐ Eliminate double handling ☐ Use efficient motions	**Reducing Fatigue** ☐ Use arm rests and other types of supports ☐ Use fixtures ☐ Add straps, handles, handholds ☐ Use power tools (light-weight) ☐ Reduce carry distances ☐ Reduce pushes / pulls ☐ Use floor mats
Reducing Awkward Postures ☐ Adjust workstation and chairs ☐ Keep items within easy reach ☐ Remove barriers ☐ Work at elbow height ☐ Use bent handle tools	**Good Standing Posture** ☐ Neck straight ☐ Shoulders relaxed ☐ Elbows at side ☐ Keep wrists in same plane as forearm ☐ Keep elbows below the shoulders ☐ Maintain the S-curve of the back	**Good Sitting Posture** ☐ Operate controls without reaching, bending, or twisting your wrists or body ☐ Seat back - 90° to 125° ☐ 2" clearance between knees and front of seat cushion ☐ Seat height 2 inches below knees when standing ☐ Relax shoulders and upper arms - position perpendicular to floor ☐ Keep arms and elbows close to body ☐ Position thighs parallel to the floor ☐ Position lower legs perpendicular to floor ☐ Rest feet firmly on floor or use footrest
Minimize Pressure Points/Contact Sress ☐ Use tools with curved handles that follow contour of hand ☐ Use tools with rounded handles – no finger grooves ☐ Add padding to sharp edges	**Improving Work Environment** ☐ Provide good, adjustable lighting ☐ Provide task lighting ☐ Place workstations perpendicular to windows ☐ Provide temperature controls ☐ Provide air conditioned break areas ☐ Provide humidity controls	

TOOL E
Hand Tool Checklist

Purpose

To evaluate and compare design features of hand tools.

When to Use It

The Hand Tool Checklist can be used prior to purchasing new hand tools or when evaluating hand tools for risk factor exposures.

How to Use It

The Hand Tool Checklist consists of a list of design criteria that are based on ergonomic principles. The checklist can be used when selecting new hand tools to ensure that they meet these design criteria and they do not result in risk factor exposures, such as pressure points, awkward postures, or excessive vibration. Comparisons can be made among new tools to assist in deciding which tool to purchase. The checklist can also be used to compare a new tool with an old tool to ensure that the new tool meets specific design features important to the task for which the tool was selected.

How to Complete the Tool

Evaluation Completed By: Add name of person completing the evaluation.

Date: Date evaluation conducted.

Task: Describe task that will be completed with tool being evaluated.

Tool 1 (Describe): Provide name of tool being evaluated.

Manufacturer: Indicate the manufacturer of the tool.

Model: Indicate the model number/name of tool.

Tool 2 (Describe): If more than one tool is being evaluated, provide name of second tool.

Manufacturer: Indicate the manufacturer of the second tool being evaluated.

Model: Indicate the model number/name of second tool being evaluated.

Questions: For each tool being evaluated, check "Yes" if the tool has or meets the design criteria described, check "No" if the tool does not have or meet the design criteria described, or check "NA" if the design criteria does not apply to the tool being evaluated.

Totals: Indicate the number of "Yes," "No," and "NA" responses for each tool. These numbers provide a quick look at how many design criteria are met by each tool.

Other Features: Indicate any other positive feature of the tool that was not included in the questions.

NOTE: The Hand Tool Checklist is based on a checklist published in the OSHA Draft Proposed Ergonomic Protection Standard, Addendum B-1, Assessment of and Solutions to Worksite Risk Factors, March 20, 1995. The checklist has been formatted to allow for a comparison among tools. The checklist is not meant to be inclusive of all design features, but to highlight major design features. While a "Yes" answer indicates a more ergonomic design, you will have to consider the function of the tool and the task it is being used to complete to determine if one of these features is more or less important.

An additional resource is *Easy Ergonomics: A Guide to Selecting Non-Powered Hand Tools* [NIOSH and Cal/OSHA 2004].

Hand Tool Checklist

Evaluation Completed By _____ Date _____

Task _____

Tool 1 (Describe) _____ Manufacturer _____ Model _____

Tool 2 (Describe) _____ Manufacturer _____ Model _____

Questions	Tool 1	Tool 2
Does the tool:		
Reduce exposure to localized vibration?	☐ Yes ☐ No ☐ NA	☐ Yes ☐ No ☐ NA
Reduce hand forces?	☐ Yes ☐ No ☐ NA	☐ Yes ☐ No ☐ NA
Reduce/eliminate bending or awkward postures of the wrist?	☐ Yes ☐ No ☐ NA	☐ Yes ☐ No ☐ NA
Avoid pinch grips?	☐ Yes ☐ No ☐ NA	☐ Yes ☐ No ☐ NA
Is tool evenly balanced?	☐ Yes ☐ No ☐ NA	☐ Yes ☐ No ☐ NA
Does tool grip/handle prevent slipping during use?	☐ Yes ☐ No ☐ NA	☐ Yes ☐ No ☐ NA
Is tool equipped with handle that:		
Does not end in palm?	☐ Yes ☐ No ☐ NA	☐ Yes ☐ No ☐ NA
Is made of textured, nonconductive material?	☐ Yes ☐ No ☐ NA	☐ Yes ☐ No ☐ NA
Has a grip diameter suitable for most workers (or are different sized handles available)?	☐ Yes ☐ No ☐ NA	☐ Yes ☐ No ☐ NA
Is made of padded or semipliable material?	☐ Yes ☐ No ☐ NA	☐ Yes ☐ No ☐ NA
Is free of ridges, flutes, or sharp edges?	☐ Yes ☐ No ☐ NA	☐ Yes ☐ No ☐ NA
Can tool be used safely with gloves?	☐ Yes ☐ No ☐ NA	☐ Yes ☐ No ☐ NA
Can tool be used by either hand?	☐ Yes ☐ No ☐ NA	☐ Yes ☐ No ☐ NA
Can trigger be operated by more than one finger to avoid fatigue?	☐ Yes ☐ No ☐ NA	☐ Yes ☐ No ☐ NA
Does tool minimize twist or shock to hand? (In particular, observe reaction of power tools due to torque.)	☐ Yes ☐ No ☐ NA	☐ Yes ☐ No ☐ NA
Total the number of Yes, No, and NA responses	__ Yes __ No __ NA	__ Yes __ No __ NA

Are there any other positive features for each tool not listed above?

Tool 1	Tool 2

TOOL F
Manual Task Risk Assessment

Purpose

To conduct a risk assessment of risk factor exposures associated with manual tasks.

When to Use It

Use the Manual Task Risk Assessment Form when you are evaluating risk factor exposures and would like to have a risk ranking that can be used to prioritize interventions, demonstrating reduction in exposures, or to focus on body parts most affected by the exposures.

How to Use It

This form can be used to evaluate risk factor exposures associated with manual tasks to rank the risk factor exposures and determine affected body parts. This information can then be used to target specific interventions and to prioritize tasks for interventions. The information can also be used to promote the success of interventions (see the Ergonomic Task Improvement Form).

How to Complete the Tool

Task: Describe the manual task being evaluated.

Date: Date evaluation is being conducted.

Assessed by: Name of person conducting the evaluation.

In consultation with: Name of person assisting with the evaluation.

Comments: Describe the task in detail, including:
- Why the task is being assessed, such as to investigate an injury report or discomfort report, evaluate productivity issues, conduct baseline assessment, etc.
- Tools – powered or nonpowered tools (include manufacturer, model, size, etc.)
- Equipment – mining equipment, lifting assist devices, transportation equipment, etc.
- Materials – any materials needed to complete the task such as equipment parts, building materials, supplies, etc.
- Overall process that includes the task being assessed.

Exposure Risk Assessment:
- Identify all primary risk factor exposures, and circle the descriptions that are most applicable.
- Circle the duration description that is most applicable to the task.
- Circle the repetition description that is most applicable to the task.
- Check any other secondary risk factors present that are listed in the box on the right side of the page.
- Determine the risk assessment score by adding the numbers listed above the circled boxes. If additional risk factors are present, add a plus sign (+) to the score for each secondary risk factor exposure.

Example:

A miner drives a haul truck for a 10-hour shift. The road is rough, and the miner experiences periodic jolting and jarring. The miner has a 30-minute lunch and two 15-minute breaks. Because of productivity requirements, the miner has 10 minutes to load, haul, and dump the product and then return for another load.

1. The risk factor exposure is whole-body vibration. Because jolting and jarring occurs, circle the box "high amplitude whole-body vibration" (score = 4).
2. Because the exposure occurs throughout the shift, circle "performed continuously for majority of shift" (score = 8).
3. Because of time pressure from productivity requirements, check "High time pressure." Also check "Lack of opportunities for social interaction," because the miner is isolated in the haul truck for most of the shift (score = ++).
4. Total score = 12++

Body Part Injury Risk: Using the risk assessment conducted in the previous section of the form, transfer the scores to the body regions affected by the risk factor exposures. Once the scores have been transferred, add up the individual scores to obtain the total score for each body region. If you have scores for more than one body region, then the body region with the highest total score would be the focus of the intervention.

Controls: This section of the form can be used to document controls that may be implemented to reduce or eliminate the risk factor exposures. Potential interventions can be identified during discussions or brainstorming sessions with miners, engineers, supervisors, etc.

NOTE: The Manual Task Risk Assessment tool was developed by Robin Burgess-Limerick, Ph.D., CPE (University of Queensland, Australia). A software version is available by contacting him at: robin@burgess-limerick.com

Manual Task Risk Assessment

TASK: LOCATION: DATE:

ASSESSED BY:

IN CONSULTATION WITH:

COMMENTS
(Reason Assessed; Tools, Equipment, Materials, Processes involved, etc.)

Manual task: Any activity requiring the worker to grasp, manipulate, strike, throw, carry, move, hold, or restrain an object, load, or body part.

Assess the degree of exposure to each primary risk factor for each body region using the table below. Determine whether any of the additional risk factors listed are present. For purposes of priority setting, a risk ranking may be determined using the numeric ratings in the table.

	Green Rating: 1	Yellow Rating: 2	Orange Rating: 4	Red Rating: 8
Exertion	Low force and speed	Moderate forces or speed, but well within capability	High force or speed, but not close to maximal	Forces or speeds close to the person's maximum
Duration	Performed infrequently for short periods	Performed regularly, but with many breaks or changes of task	Performed frequently, without many breaks or changes of task	Performed continuously for most of shift
Repetition	Dynamic and varied patterns of movement	Little or no movement, or repeated similar movements	Repeated identical movements	
Posture	Comfortable postures, within a normal range about neutral	Uncomfortable postures, but not involving postures at the extreme of the range of motion	Postures at the extreme of the range of motion	
Vibration	No hand-arm or whole-body vibration	Moderate-amplitude hand-arm vibration or whole-body vibration	High-amplitude hand-arm vibration or whole-body vibration	

☐ Hot or cold environment
☐ High stress environment
☐ High time pressure
☐ Lack of control over work
☐ Cognitive over/under load
☐ Lack of opportunities for social interaction

Determine the body region(s) that may be at risk of injury. (Alternatively, assess the task for each of the following regions: lower limbs; lower back; neck/shoulders and upper back; elbows, wrists, and hands.)

Body Region	Exertion	Duration	Repetition	Posture	Vibration	Total Risk Score
Neck, Shoulders, and Upper Back						
Elbows, Wrists, and Hands						
Lower Back						
Legs, Knees, and Feet						

Engineering Controls	Administrative Controls	Personal Protective Equipment

TOOL G
Ergonomic Task Improvement Form

Purpose

To provide an effective method to highlight interventions implemented to reduce or eliminate risk factor exposures.

When to Use It

Use the Ergonomic Task Improvement Form after an intervention has been implemented.

How to Use It

This form can be use to compare how a task was done before and after an intervention was implemented to demonstrate the reduction of the risk factor exposures. Forms can be posted on bulletin boards or the company Intranet. If similar operations are conducted at multiple sites, the form can be used to provide ideas for interventions at other sites. An example of a completed form is shown below.

How to Complete the Tool

Name of Manual Task: Add name of task in the title (in the example below, "Moving Electromagnetic").

Division: Add name of division.

Mine: Add name of mine.

Department: Add name of department.

Task Description: Briefly describe the task, including information pertinent to the risk factor exposures.

Equipment/Tools Used in Task: List the specific equipment and tools involved.

Frequency of Task: Indicate how many times the task is done.

Number of Workers Affected: Indicate the number of workers who perform this task.

Employee Concerns: Describe the number and type of injuries, or the presence of body discomfort reported by employees associated with doing this task.

Risk Assessment: Complete the table based on risk assessment results.

Objective of Control Measure: Indicate how the risk factor exposures will be reduced or eliminated.

Description of Control: Briefly describe the controls used to reduce or eliminate the risk factor exposures. Include model number if applicable.

Manufacturer/Contact Information: Provide the source of the control measure, including contact information (phone and/or e-mail address).

Cost: Provide cost information for the control. If the control was constructed in-house, provide an estimate of materials and labor costs.

Effect of Control on Productivity: Indicate if the control resulted in either an increase or decrease in productivity.

Effect of Control on Injury Risk: Indicate if the control is expected to impact injury risk.

Risk Assessment: Repeat the risk assessment after the implementation of the control.

Further Actions / Administrative Controls Required: Include information on actions that will be needed as a result of implementing the control, such as maintenance, inspections, reports, schedule changes, etc.

NOTE: If possible, include photographs to show how the task was done before and after implementing the control, and include the worker in the photographs. Follow the same color scheme in the risk assessment tables that was defined in the Manual Task Risk Assessment.

Ergonomic Task Improvement
Task: Moving Electromagnet

DIVISION_____ MINE_____ DEPARTMENT_____

TASK DESCRIPTION: The electromagnet is manually pulled to the scrap bin by pulling down on a chain over a pulley. The magnet is heavy, and maximum effort is required. The worker reaches to his maximum height to grasp the chain and pull down using his entire body weight.

EQUIPMENT/TOOLS USED IN TASK: Electromagnet

FREQUENCY OF TASK: Daily   **NUMBER OF WORKERS AFFECTED:** 5

ROOT CAUSES OF RISK FACTORS: The mass of the magnet requires very high force to move.

EMPLOYEE CONCERNS: An acute shoulder injury was reported and resulted in a lost-time injury.

BEFORE

Body Region	Exertion	Duration	Repetition	Posture	Vibration	RISK RANK
Upper Body: Neck, Shoulders, & Upper Back	8	1	4	4	1	18
Upper Limb: Elbow, Wrist, Arm, and Hand	8	1	4	1	1	15
Lower Back	2	1	2	1	1	7
Lower Limb: Leg, Knee, and Foot	2	1	1	1	1	6

5-10 = Low Risk   11-15 = Medium Risk   16-24 = High Risk

OBJECTIVE OF CONTROL MEASURE: Substitute the manual effort with a winch that pulls the magnet into place above the scrap bin.

DESCRIPTION OF CONTROL (tool, equipment, or work station changes/purchases): Electric winch purchased and installed by contractor.

MANUFACTURER/CONTACT INFO: Acme winches

COST: $5,000

EFFECT OF CONTROL ON PRODUCTIVITY: Winch is considerably faster.

EFFECT OF CONTROL ON INJURY RISKS: Risk of injury eliminated.

AFTER

Body Region	Exertion	Duration	Repetition	Posture	Vibration	RISK RANK
Upper Body: Neck, Shoulders, & Upper Back	1	1	4	4	1	6
Upper Limb: Elbow, Wrist, Arm, and Hand	1	1	1	1	1	5
Lower Back	1	1	1	1	1	5
Lower Limb: Leg, Knee, and Foot	1	1	2	1	1	6

FURTHER ACTIONS/ADMINISTRATIVE CONTROLS REQUIRED: Winch maintenance needs to be added to maintenance schedule and inspected.

TOOL H
Risk Factor Cards

Purpose

The risk factor cards were designed for use as a guide to identify risk factor exposures and to suggest some simple job improvements.

When to Use It

Examples of the cards are shown in Figure 10. The primary use of the cards is as a handout following training on risk factor exposures. The cards serve as a reminder to employees to identify and report risk factor exposures associated with their jobs. The cards are slightly larger than a credit card, and can be easily carried in the pocket. Information is printed on both sides of the cards. For cards highlighting risk factors, examples of risk factor exposures are provided on the front of the card, with potential solutions to the risk factors provided on the reverse.

Because electronic versions of the cards are provided, poster-sized versions can be made and placed at locations where employees would have an opportunity to view them, such as a break room or locker room. Posters can be displayed in conjunction with toolbox or refresher training addressing risk factor exposures. The cards can also be used by trainers as a handout during training sessions to reinforce information presented during the training. The cards can be either laminated or printed on plastic sheets to improve durability.

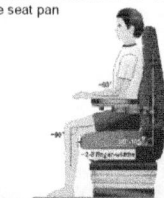

Figure 10.—Examples of risk factor cards. (The front of each card is shown on the left, the reverse on the right.)

TOOL I

Ergonomics Sticker

Purpose

The stickers are meant to reinforce the application of ergonomics to improve mining jobs.

When to Use It

The sticker, shown in Figure 11, is primarily intended as a handout following training on risk factor exposures. It serves as a reminder to participate in the ergonomics process by identifying and reporting risk factor exposures associated with jobs. The sticker can be placed on hardhats, lunch boxes, locker doors, etc. It can also be used as an incentive to encourage employees to report risk factor exposures. Employees would receive a sticker after reporting a potential exposure.

The sticker can be modified to make it specific to a particular company. Figure 12 shows the sticker as modified by Bridger Coal Co. (name of company added to sticker) and Vulcan Materials Co. (colors changed to company colors).

Figure 11.—Surface mining sticker promoting ergonomics.

Figure 12.—Stickers modified by Bridger Coal Co. *(left)* and Vulcan Materials Co. *(right)*.

Section V

Training

Introduction

According to Cohen et al. [1997], training is one of seven critical elements of the pathway to controlling MSDs. Training is important because it ensures a basic level of knowledge necessary for individuals to effectively fulfill their role as participants in an ergonomics process. As with any safety and health training, it is important to address training needs at multiple levels and functions, as shown in Table 13. This table includes suggested training topics specific to the different groups typically involved with implementing an ergonomics process.

Although basic ergonomics training is commercially available, it may not necessarily address tasks specific to mining or consider some of the challenges in controlling risk factor exposures, such as the dynamic nature of mining tasks, harsh environmental conditions, and restricted work spaces. To assist mining companies with providing training to employees, NIOSH developed a train-the-trainer package. A description of this training (Ergonomics and Risk Factor Awareness Training for Miners) is included in this section and is available as a separate NIOSH publication (DHHS (NIOSH) Publication No. 2008–111, Information Circular (IC) 9497). Training to inform management about the benefits of an ergonomics process was also developed by NIOSH. A description of this training and the actual presentation are included in this section. Training for BBS observers is discussed in this section as well. The content for this training should be specific to the mine site, so only an outline and method of conducting this training are provided. The employee training package and the management training are available on the NIOSH Mining Web site (www.cdc.gov/niosh/mining).

Table 13.—Suggested training topics for groups involved with implementing an ergonomics process

Groups	Training topics							
	Value of Ergonomics Process	Resources Needed	Examples of Successful Processes	Basic Ergonomics Awareness	MSD Risk Factor Identification	MSD Risk Factor Evaluation	MSD Risk Factor Controls - Basic	MSD Risk Factor Controls - Advanced	Problem-solving and Team Dynamics
Management	X	X	X						
Supervisors	X	X	X	X	X		X		
Employees				X	X		X		
Ergonomics Team / Committee	X	X	X	X	X	X		X	X
Engineering				X	X			X	
Purchasing				X	X		X		
Medical				X	X		X		
Safety and Health	X	X	X	X	X	X		X	
BBS observers				X	X		X		

Ergonomics and Mining: Ensuring a Safer Workplace – Training for Management

When implementing an ergonomics process, several basic elements have been identified in order for the process to be successful. One of these elements is management support. This support determines the goals, resource levels, communication levels, and process evaluations for continuous improvement. When management support is observable, it denotes the importance of the process to the organization and its employees. Because management may not know how ergonomics can be applied within the organization and how it can be beneficial, it is often necessary to provide this information to obtain their buy-in for the ergonomics process.

Objectives

The objectives are to provide management with an understanding of how ergonomics processes can add value to the organization (primarily by reducing MSDs) and to obtain their support for implementing a process.

Content/Topics

This training is designed as an initial step to obtain management support for implementing an ergonomics process. The training should be given by safety and health managers or other personnel responsible for ergonomics. The training is a Microsoft Powerpoint file and can be modified to include information specific to your company. This file does not include any background or slide design formatting, so it can be modified easily to suit your particular needs. It takes approximately 90 minutes to present; however, the training can be shortened. Four examples of successful programs are included in the training; one or two of these examples could be omitted if the presentation time needs to be reduced. Notes are also provided to assist in presenting each slide.

The training addresses the following topics:
- What is Ergonomics?
- Costs of MSD Injuries
- Ergonomics as a Solution
- Successful Programs at Other Companies
- Goals and Next Steps – This topic can be used to explain the short- and long-term goals, what is needed to implement the ergonomics process, and how the implementation will be accomplished.

This training is on the CD included with this document and is also downloadable from the NIOSH Mining Web site (www.cdc.gov/niosh/mining).

Ergonomics and Mining: Ensuring a Safer Workplace

Management Training

(Provide an explanation of why this presentation is being given to your management. Additional slides highlighting specific problems associated with exposures to MSD risk factors within your organization can be added to this presentation.)

Presentation Outline

- What is Ergonomics?
- Costs of MSD Injuries
- Ergonomics as a Solution
- Successful Programs at Other Companies
- Next steps – **Where we are headed**

Ergonomics - What is it?

Most people look like this...

Some designers must think people look like this...

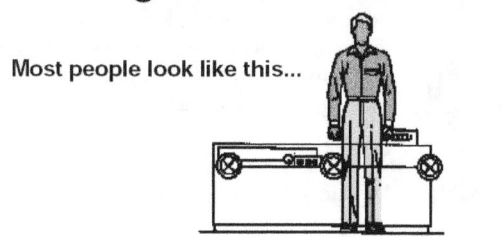

You can see here that the way equipment is designed does not always consider the worker using it. Workers are often put in situations where their tools and equipment are designed so that they must get into awkward positions (reaching and stooping) in order to operate the equipment.

Ergonomics is...

- Scientific study of human work.
- Considers physical and mental capabilities of workers as they interact with tools, equipment, work methods, tasks, and working environment.
- Goal – to reduce work-related injuries by adapting work to fit people *instead of* forcing people to adapt to work.

The idea behind ergonomics is to adapt the workplace to fit the workers. To do this, we need to use what we know about the limitations and capabilities of people. Then we need to make sure the work does not require people to work outside of these capabilities or stretch them beyond their limits. When the work is too much, we need to change the work station, the tools and equipment people use, or the way that people do their jobs.

Costs of Musculoskeletal Disorders

National Academy of Sciences Study

- **1 million workers miss time from job each year**
    - Upper-extremity and low back disorders
- **$50 billion in direct costs**
- **$1 trillion if you include indirect costs**
    - 10% U.S. Gross Domestic Product
    - Reduced productivity, loss of customers due to errors made by replacement workers, and regulatory compliance

When work requirements exceed the limitations and capabilities of workers, injuries occur. For instance, each year 1 million workers miss time from their jobs because of MSDs. This costs over $1 trillion when both direct and indirect costs are considered.

(Source: National Research Council [1999])

Costs of Musculoskeletal Disorders

- **Median number of lost work days**
    - 5 days for all workers
    - 25 days for workers with MSDs
- **Average cost per injury (UE)**
    - $824 for all other cases
    - $8,070 for an MSD
- **MSDs tend to have**
    - Longer durations
    - Longer treatment time
    - Greater work disability

These statistics give you an idea of the breakdown of MSD costs during the 1990s. In general, MSDs result in more lost days and cost 10 times more than other types of injuries. MSDs result in longer duration, longer treatment times, and greater disability. Because of the aging workforce, these costs are expected to get worse.

Sources:
Days lost data: U.S. Bureau of Labor Statistics [1998].
Cost data: Webster and Snook [1994].
MSD info (1993–1994): Feuerstein et al. [1998].

MSDs in Mining

- Illnesses reported to MSHA:

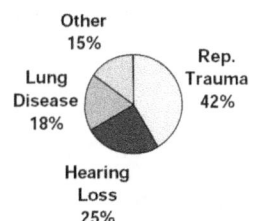

- Non-fatal lost-time injuries
  - 33% handling materials
  - 16% slip/fall
  - 16% fall of ground
  - 12% powered haulage
  - 10% machinery

While MSDs have been a problem in mining for a long time, they have not been at the forefront, partially because they are not regulated. Unlike dust, noise, and diesel issues, MSDs are not regulated by standards. Consequently, they are not the first priority at most mines.

However, MSDs account for over 40% of injuries and illnesses, significantly affecting the bottom line of companies either directly or indirectly. For that reason, it has been possible to gain buy-in from some mines to reduce MSDs. More and more mine companies realize that this is an issue that is no longer optional if they want to significantly reduce their injury rates. With the rising costs of health care and the difficulty in getting and keeping quality mine workers, they want to make changes to the work environment to reduce MSDs.

Age and MSDs

- Median Age for miner is 42.2 years
- 40% of all injuries are MSDs for miners age 35-55
- Older workers incur approximately 3 times as many days lost than younger workers

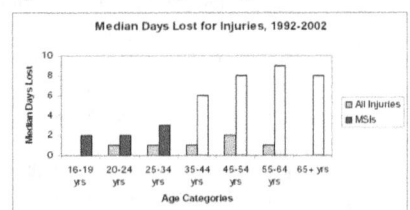

Age is associated with higher rates of MSDs

Compounding the problem of MSDs is the issue of the aging workforce and the effect of MSDs on this workforce. The median age for miners is 42.2 years. For miners between 35 and 55 years old, 40% of all injuries are MSDs. In addition, older workers incur three times the number of days lost compared to younger workers.

PREVENTION: to invest, or NOT to invest...

40 Lost Workday Mishaps in 1993

PREVENTION	$ INVESTED	LOST WORKDAY INJURIES
Safety Glasses	$ 70,000	3 F/O Eye
Safety Shoes	$120,000	2 Foot
Respirators (Med/Trn/Matls)	$124,000	1 Chemical Exposure
Ergonomics	$0	34 Strains

Why don't we budget to "prevent" the #1 injury category?

This is an example from a Florida Navy facility. Of 40 lost workday mishaps in 1993, there were 3 eye injuries from foreign objects and irritation, 2 foot injuries (stepping on a nail and being struck by a sledgehammer on the metatarsal), and 1 chemical allergy. The facility had invested $314,000 in three prevention programs (safety glasses, safety shoes, and respirators) to help prevent these types of injuries, and they still had six injuries. However, they also experienced 34 strains that year and had not invested 1 cent in ergonomics at that point. The question that needs to be answered is: Why did they not budget to prevent their No. 1 injury? (Discuss if this same situation applies to your organization.)

(Source: Wright B [2002])

Case Study – Back Injuries can Be Serious

- Supervisor suffered back injury helping worker move sheet metal in Jan '78
  - $1000 medical costs and no lost time
  - Recurrence in '92 cost **$18,000**
  - Surgery/comp in '93 cost **$81,000** and resulted in permanent partial disability vs. retirement
  - Indirect/Chargeback costs **$55,000** in '01

This <u>one</u> 1978 back injury has cost over <u>*$517,000*</u> so far!
Prognosis: not positive!

> Original injury cost does not appear to warrant investment . . . until you consider future costs.

This is another example from a Navy facility. A supervisor suffered a back injury in 1978. At that time, the costs were only $1,000 and there was no lost time. However, a recurrence in 1992 cost $18,000. In 1993, another $81,000 was spent on surgery and compensation. By 2004, the costs associated with that initial injury had risen to over $500,000. If that initial injury had been prevented, the cost savings would have been significant to the organization.

Costs are real (extended) FECA WMSD costs.

(Source: Wright B [2002])

Prevention PROACTIVE ⇒ Early Intervention

- **Complex Problem**
  - Do <u>not</u> know a lot about MSDs and how they occur
  - Do <u>not</u> know why some individuals are more susceptible than others
  - Do <u>not</u> know what are safe levels of exposure
  - Do know risk factors associated with MSDs

- Well-designed workplace interventions prevent MSDs and early medical care reduces severity.

THE EARLIER ACTION IS TAKEN, THE GREATER LIKELIHOOD OF SUCCESS!

A "one size fits all mentality" does not apply to ergonomics. There is a lot we do not know about MSDs, such as:
- How they occur
- Why some people are more likely to get an MSD compared to others
- What are safe levels of exposures to risk factors associated with MSDs

However, we do know the risk factors that result in MSDs, and we know that well-designed work tasks and interventions prevent MSDs.

If an MSD does occur, we also know that early medical care reduces the severity of the MSD.

Targeting Risk Factors

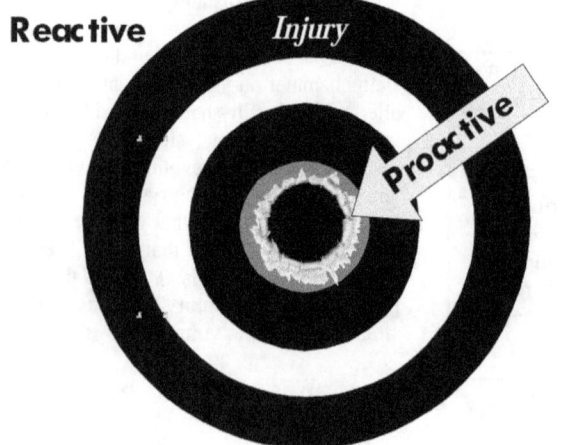

We want to change our way of thinking about injuries. We want to stop being reactive and start being *proactive*. To do this, we must understand injuries, the signals that tell us when one might occur, and how to prevent them from happening.

We want to avoid waiting for injuries to occur before we take action.

We want to be able to identify the risk factors for job tasks and equipment used and then take correction action before an MSD or even discomfort occurs.

Risk Factor Examples

Examples of Forceful Work
- Heavy lifting
- Carrying heavy objects
- Forceful pushing or pulling
- Forceful gripping
- Shoveling damp or heavy materials

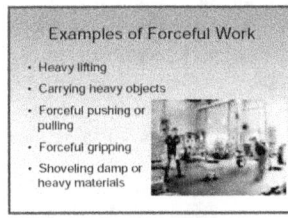

Examples of Poor Posture
- Elbows above shoulders
- Extended forward reaches
- Trunk bent over more than 20 degrees
- Twisting the trunk
- Extreme wrist bending
- Pinch grips
- Kneeling or squatting

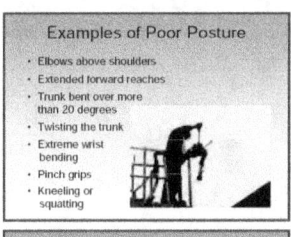

Examples of Repetitive Work
- Using equipment controls
- Machine paced assembly tasks
- Packing or unpacking items
- Computer keyboarding
- Manning a store checkout line

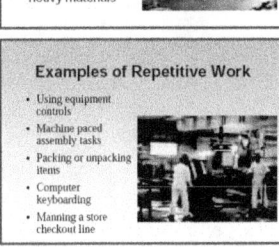

Examples of Vibration Exposure
- Hand-Arm: Using vibrating tools.
- Whole-Body: Sitting or standing on vibrating surfaces. (Includes jolting & jarring)

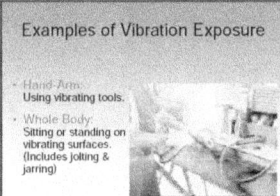

These are examples of common risk factors and some tasks during which the risk factor exposures occur. The four main risk factors are: forceful work, poor posture, repetitive work, and vibration.

Why Target Risk Factors?

The cumulative nature of musculoskeletal disorders:
...an exponential relationship.

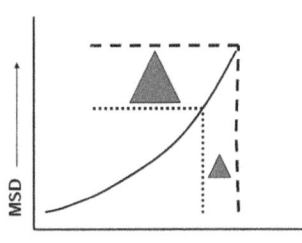

The relationship between risk factor exposure and the resulting MSD is exponential. The risk factor exposure usually occurs over a long period of time before an MSD occurs. When the MSD occurs, we often discover the worker had been doing a task that he or she had done many times before without getting injured. It is perplexing as to why the MSD happened. However, because the effect on the worker is gradual, he or she may not recognize that a problem is occurring. When the MSD does occur, it only takes a minimum increase in exposure to result in a MSD, often with lost time resulting.

An Ergonomics Process . . .

What is it?

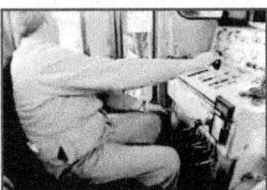
Drill Operator

- A systematic method to improve the fit between the worker and the workplace to improve safety, productivity and workforce satisfaction.
- Can be a stand alone process or it can be integrated with existing safety and health programs
- Works best as a participatory process – management and employees

PROACTIVE APPROACH TO PREVENTION

An ergonomics process is a systematic method to improve the fit between workers and their workplace (including work stations, equipment, tools, and environment) to improve safety, productivity, and workplace satisfaction. An ergonomics process can be a stand-alone process or it can be integrated with existing programs, such as safety and health. It can be viewed as a third prong of a comprehensive safety and health program: safety, industrial hygiene, and ergonomics. Ergonomics processes usually work best when employees participate fully and when there is a champion to keep the process moving forward. A champion is usually needed until the process becomes embedded in the organization and culture.

Injury/Illness Prevention Process

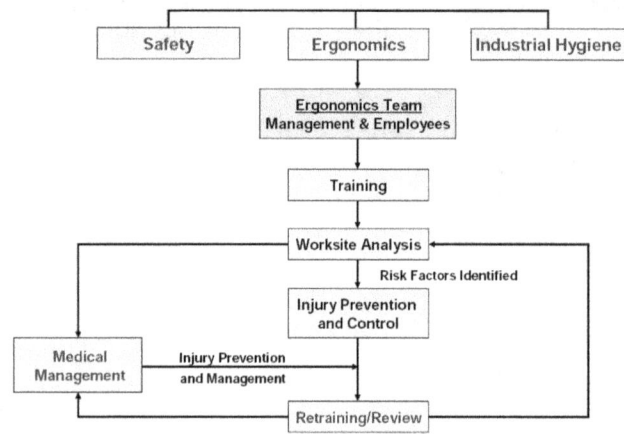

This is an example of a simple process. Form a team, train the team members, conduct a worksite analysis to identify risk factor exposures, implement controls to reduce or eliminate exposures, provide medical management for workers with injuries, and finally, review or conduct a self-assessment of the process and provide retraining if needed.

Successful Program Examples

- GAO Report – Ergonomics programs:
  - Reduced compensation costs 35%–91% in five diverse companies
  - Increased safety and health
  - Increased efficiency of operations
  - Increased profitability
  - Increased quality of life
- American Electric Power (AEP)
- CONSOL
- Jim Bridger Mine, Bridger Coal Co.

We have good reason to suggest the application of ergonomic principles as a viable solution to MSDs. The effectiveness of ergonomics can be found in a GAO report published in 1997, along with some examples from mining companies.

Government Accounting Office (GAO)

American Express (5,300)
AMP, Incorporated (300)
Navistar International Transportation Corp. (4,000)
Sisters of Charity Health System (780)
Texas Instruments (2,800)

The first examples of successful processes were published by the GAO in 1997. They reported on ergonomics processes implemented in five different companies, varying in size (from 300 to 5,300 employees) and industry.

GAO Results

Percentage Reduction in Workers' Compensation Costs for MSDs

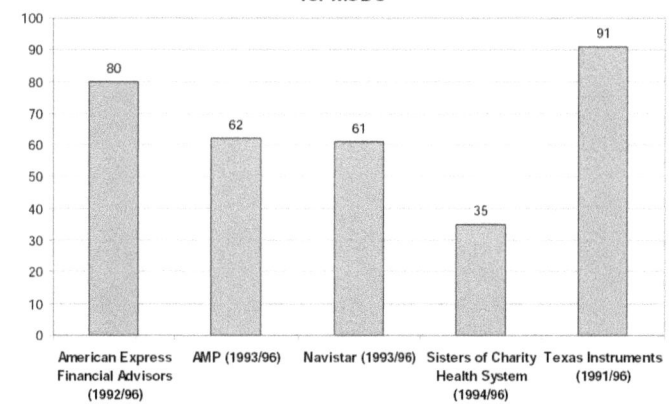

This bar graph shows the percentage reduction in workers' compensation costs for MSDs at each of the five companies included in the GAO report. The range was a 36–91% reduction in compensation costs over a 2 to 5-year time period.

GAO Results

Average Dollar Cost per MSD Workers' Compenstion Claims

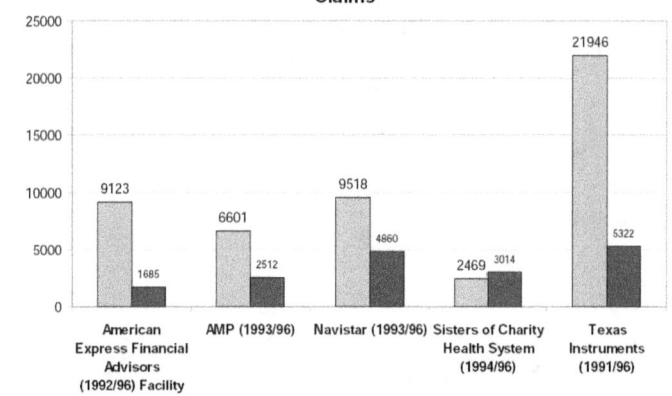

Each company, except one, experienced a decline in the average dollar cost per MSD. The one company that did not show a decline also had the youngest process (only 2 years). It is likely this would change as its process became more mature.

GAO Summary

- 36% to 91% reduction in worker's comp costs, overall reduction in lost workdays
- Core elements: management commitment, employee participation, ID of problem jobs, analysis and improvements, training and education, and medical management.
- Inexpensive, "low tech" improvements are effective
- Incident-based vs. risk factor-based

In summary, the companies included in the GAO report:

- Experienced a 36–91% decrease in workers' compensation costs
- Followed the core elements mentioned earlier
- Showed that inexpensive improvements were effective in reducing exposures

However, all of the processes were incident-based (reactive) rather than risk factor-based (proactive).

(Source: GAO [1997])

American Electric Power

The second example of an ergonomics process is from American Electric Power. This is a mining example.

AEP's Approach

- Goal:
  To reduce MSD problems at coal mining operations with initial emphasis on back injuries

- 1989 - set standards and objectives to establish ergonomic committees at its six mining operations

- Provided charter for structure and function of committees

- Established method for followup of committee activities, ensuring a proper communication feedback loop

- Provided corporate support for all technical and financial needs and training for committee members

The initial goal of AEP's process was to reduce MSDs of the back. To accomplish this goal, AEP established Ergonomics Committees, supported by a defined structure, specific functions, an effective communication process, and adequate training and resources.

AEP Ergonomics Committee

Participants   Activities

Ergonomics Subcommittee

Activities

AEP - Committee Members

- Management
  - Mine superintendent/Chairman
  - General mine supervisor
  - Maintenance supervisor
  - Production supervisors
  - Shift supervisors
  - Safety manager
- Technical
  - Belt coordinators
  - Supply coordinators
- Labor
  - UMWA safety committee chairman
  - Two hourly employees (rotating)

The membership of the Ergonomics Committee included management, labor, and technical positions.

AEP - Committee Activities

- Do front-end needs analyses
- Develop problem statement & solutions
- Develop plans, standards, & proposals
- Monitor implementation
- Provide counsel & feedback
- Provide regular reports to corporate management

This is the process used by AEP's Ergonomics Committees. They studied the issue, developed and implemented solutions, monitored effectiveness, and then provided feedback. The committees also reported regularly to corporate management.

AEP Subcommittee Activities

- Conduct worker surveys
- Audit work activities
- Conduct observations of work activities
- Develop improvement projects
- Submit improvement projects to senior committee for approval

Subcommittees were formed to focus on a target work area. The subcommittees were chaired by the department head and consisted of a number of workers from the target area. They met at least once a month and worked with their Ergonomics Committee to set goals and objectives.

Program Assessment (1991)

- Most members were satisfied with involvement on committees, although some believed they should be devoting more time.
- Most members thought it improved their ability to do their jobs.
- Many were dissatisfied with the amount of time it took for suggestions to get implemented.

An assessment of the process in 1991 indicated that most members were satisfied with being involved with the process and thought it improved their ability to do their jobs. Many of the members thought it took too long to implement changes.

AEP Successes - Ergonomic Applications

- Implementation of standard hoist mechanisms eliminated handling of heavy materials
- Reducing object weights (either bagged or stopping materials) reduced back strain injuries – work with suppliers
- Proper storage of wood products (reducing exposure to water) prevented additional weight to be handled

Some of the job improvements implemented included:
- Using standard hoist mechanisms to lift heavy materials
- Reducing weight of supplies (bagged or stopping materials)
- Properly storing wood products to eliminate exposure to water, which increased the weight of the wood

AEP Successes - Specialized Mining Equipment

- In-house tools designed for specific mining applications, such as a tool used to remove or install conveyor belt rollers
- Zipmobile – a materials-handling cart that moves supplies along the longwall face
- Belt car allows miners to splice 500 feet of belt without manual handling
- Shuttle cars have air-ride suspension to reduce whole-body vibration
- Ergobus moves tools and equipment for maintenance and outby tasks

Other interventions implemented by AEP included various specialized mining equipment, such as:
- Tool to remove or install conveyor belt rollers
- Zipmobile to move supplies along the longwall face
- Belt car that eliminated manual handling
- Shuttle cars with air-ride suspension to reduce whole-body vibration
- Ergobus to transport maintenance tools and equipment

AEP Successes - Back Injuries

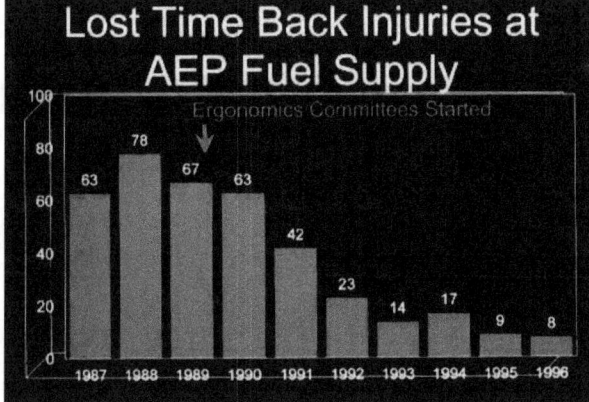

As a result of its efforts, AEP saw a significant reduction in lost-time back injuries. In 1988, the number of back injuries was 78. By 1996, the number of back injuries had dropped to 8 (a 90% decrease).

AEP Recommendations

- Diverse committee members (no more than 12)
- Team players
- Select committed, open-minded leader with skills to run effective meetings
- Training on group problem-solving
- Organize campaign to describe purpose of team
- Committee must be backed by entire organization

From their experience, AEP made these recommendations regarding implementation of an ergonomics process. They emphasize the need to select the right individuals for the committee, to train the committee members, and then to ensure support from the entire organization.

Comment from AEP

Tim Martin, Safety and Health Manager
1997 Ergonomics Conference, Chicago, IL

"What exactly have we accomplished with our ergonomics programs?

We definitely have reduced our accidents, reduced our compensation costs drastically, increased productivity, reduced down time, increased a lot of employee involvement, and our relations with our employees have really grown quite drastically…We feel that without ergonomics programs, we couldn't have accomplished this."

Tim Martin, the safety and health manager for AEP, had this to say about their ergonomics program:

"What exactly have we accomplished with our ergonomics programs? We definitely have reduced our accidents, reduced our compensation costs drastically, increased productivity, reduced down time, increased a lot of employee involvement, and our relations with our employees have really grown quite drastically…We feel that without ergonomics programs, we couldn't have accomplished this."

(Source: NIOSH [1997a].)

The Bridger Coal Story…

- Western surface mine
- Skeptical but proactive
- Safest and healthiest workforce
- Efficiently developed
- Highly effective
- Integrated into H&S program

The last example of implementing an ergonomics process is about the Bridger Coal Company. Bridger operates the Jim Bridger Mine, a surface coal mine in southwestern Wyoming. Bridger wanted to improve its safety and health program by integrating an ergonomics process with its existing safety and health program. Bridger attributed improvements in its health and safety culture to the ergonomics process. Although Bridger was skeptical in the beginning, the company thought proactively and wanted a more formal program.

Kean Johnson, Ergonomics Process Coordinator, stated:

"Ergonomics has played an important role in helping Bridger Coal reach our goal of providing the safest and healthiest working environment possible for our employees. Our management and hourly employees alike understand the value of what has been developed. In the beginning, when the idea of establishing such a program surfaced, we were all skeptical of just how things would work. However, thanks to the combined efforts of NIOSH, PacifiCorp, and those at Bridger Coal Company involved in the creation process, we found that an ergonomics program could not only be efficiently developed, but that it could be highly effective as well. The Ergonomics Program is currently an integral part of our company, and we are confident that it will continue to improve and enhance the safe working experience at our mine."

(For more information about the ergonomics process implemented by the Bridger Coal Co., see: Torma-Krajewski et al. [2006].)

Bridger established a multidisciplinary team to ensure they addressed all aspects of the company. They realized that each department would be affected by this endeavor.

Ergonomics Awareness Training

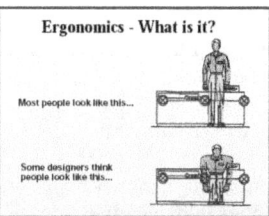
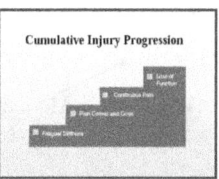
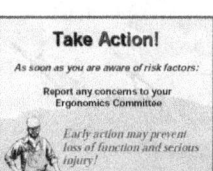

All employees received ergonomics and risk factor awareness training. Examples of risk factor exposures specific to their mine were used so that employees would not only relate to the examples, but would begin the solution process. This training focused on:

- Defining ergonomics and its benefits for employees
- Describing how MSDs develop
- Identifying risk factor exposures
- Reporting exposures

Reporting Concerns

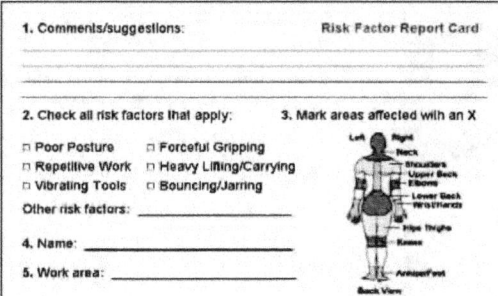

Employees learned about a simple way to record their concerns. They practiced during the training and then filled out Risk Factor Report Cards about their exposures when they went back to their jobs.

Communications & Recognition

- Bulletin board
- Posters
- Stickers
- Safety newsletter
- Audit results

One of the most important aspects of the process was communications and recognition. As employees turned in Risk Factor Report Cards and the concerns were prioritized by the Ergonomics Committee, Bridger posted information telling employees the status of their concerns. The stickers and posters reminded them of what they had learned and that the program was not a fly-by-night effort. Bridger's efforts were acknowledged in newsletters that were published and distributed throughout the company.

Interventions

- 55 concerns reported
  - 22 completed interventions
  - 5 being addressed
  - 9 on hold pending receipt of additional info
  - 19 addressed as S&H concerns
- Types of interventions
  - New equipment
  - Workstations rearranged or adjusted
  - Retrofits

Within just over a year, 55 concern cards were submitted. A few of them were collaborative efforts between the Ergonomics and Safety Committees. More than 22 successful (accepted and used) interventions were implemented.

The interventions were mostly engineering controls, and many included the purchase of new equipment. The equipment was purchased with improved knowledge of ergonomic principles, resulting in specifications that better met the needs of the end users. Bridger also redesigned or reorganized work stations, and they retrofitted specific controls for existing equipment. Very few of the interventions exceeded $1,000.

Bridger Intervention Successes

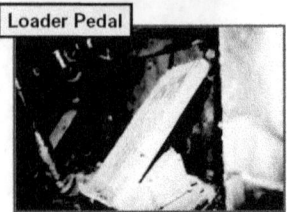

Here are a few of their inexpensive changes:

Chock blocks: Handle added that eliminated bending the back when placing or removing the chocks.

Welding helmets: Lighter-weight welding helmet eliminated neck discomfort.

Loader pedal: Angle of pedal was reduced, which eliminated foot and leg discomfort.

Water pump switch location: Switch was brought closer to the operator, which reduced the reach and shoulder discomfort.

Bridger Intervention Successes

Prill Truck Ladder

Hand rails not properly located

Hand rails moved closer to ladder

Another intervention involved moving handrails. The handrails were located too far from the ladder, and workers did not use them. If the workers attempted to use the handrails, their body would lean backward and they would have to forcefully hold onto the handrails to prevent from falling. After the handrails were moved closer to the ladder, the workers could easily and safely use the handrails.

Bridger Intervention Successes

Moving Dragline Cable

Moving dragline cable was one of the most physically demanding jobs performed. A J-hook bar is now used to drag the cable along the ground near the dragline. This eliminated the need to manually lift the cable. Away from the dragline, no physical effort is needed as the job now uses a cable-handling device attached to a front-end loader.

Bridger Intervention Successes

Dragline Armrests

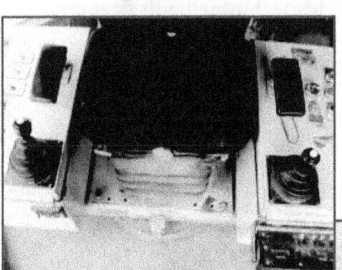

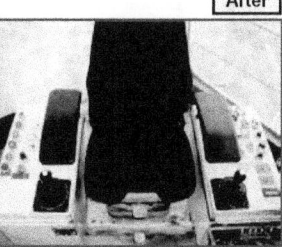

Dragline operators were complaining of upper-extremity discomfort and pain. It was determined that the support and function of the manufacturer-installed armrests were inadequate and nonadjustable. The new armrests were adjustable and provided larger nonfrictional support.

On their own....

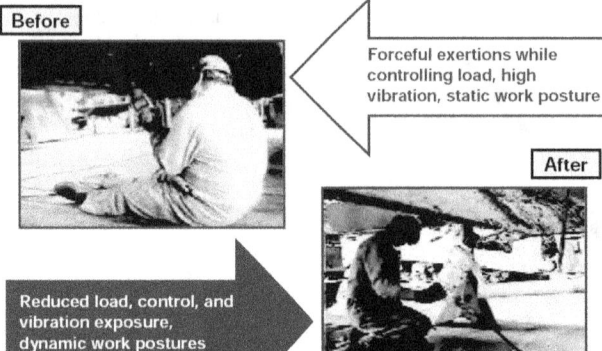

Before — Forceful exertions while controlling load, high vibration, static work posture

After — Reduced load, control, and vibration exposure, dynamic work postures

Bridger employees are doing it on their own. They developed a new way to support the 1½-inch pneumatic wrench without any forceful exertions.

Integrating Ergonomics

- Safety & health
- Corporate-wide risk assessment
- Injury investigations
- Design specifications
- Purchasing decisions

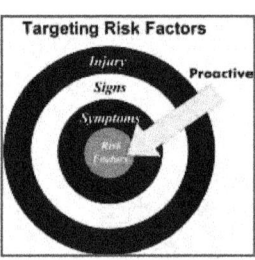

Targeting Risk Factors

To be proactive means integrating and applying ergonomic principles with many other processes to prevent exposures to risk factors. A better-educated employee makes better purchases.

Lessons Learned

- Committee Participants
- Process Development
- Process Implementation
- Supervisory Training

Lessons learned from the Bridger process included:
- Committee participants need to be interested and committed to the process, and they must be given resources.
- Each organization must decide the best way to implement its process. The "one size fits all" approach does not work in this case.
- Supervisors need to receive training regarding their role in the process.

Characteristics for Success

- **Commitment** - Strong in-house direction & support
- **Training** - Staff expertise in team building & ergonomics
- **Team Composition** - Diverse but manageable size
- **Input** – Employee input to help define team objectives
- **Communication** - Everyone kept informed of objectives, progress, & accomplishments

Simply put, successful ergonomic processes have several common characteristics: strong commitment, training, diverse team members, employee input, and communication.

What Successful Programs Can Provide

- Avoidance of illness and injury risks
- Lower worker compensation costs
- Higher productivity
- Increased workforce job satisfaction
- More employee involvement in lower-level decision-making

As demonstrated by the examples discussed during this presentation, several benefits shown here can be derived from an ergonomics process.

Final Comments – Ergonomics Processes

"Saturate the organizations with knowledge…Give the process time to work…Keep in mind that it's a cultural change, a change in the way you think and not just another program." (Tom Albin, Ergonomist, 3M)

Tom Albin of 3M made this comment about implementing an ergonomics process:

"Saturate the organizations with knowledge…Give the process time to work…Keep in mind that it's a cultural change, a change in the way you think and not just another program."

(Source: NIOSH [1997b].)

Short- and Long-term Goals

OUR SHORT-TERM GOAL
(Add your short-term goal.)

OUR LONG-TERM GOAL
(Add your long-term goal.)

(Add your short- and long-term goals on this slide.)

(Examples:)
Form an ergonomics team and provide team members with adequate training and resources.

Lower risk factor exposures associated with: (add specific job task).

Next Steps…

(Add your own next steps according to your plan.)

Ergonomics and Risk Factor Awareness Training for Miners

Research has shown that ergonomics is most successful when it is approached as a participatory process—management and employees working together to modify job tasks, including equipment, tools, environment, and methods. The first step toward achieving a participatory process is to give employees knowledge of ergonomics and how it can be used to align their jobs to their abilities.

The Ergonomics and Risk Factor Awareness Training is designed to encourage employees to report risk factor exposures and body discomfort. It also encourages employees, when possible, to make changes to their jobs to reduce their exposures to risk factors.

Objectives

The overall objective of Ergonomics and Risk Factor Awareness Training is to help reduce injuries and illnesses resulting from exposures to risk factors. Specifically, this training will increase awareness of risk factors and encourage miners to take action to report and reduce their exposures to risk factors.

Content

This training is designed specifically for the mining industry. Because mining is a diversified industry, examples representing both surface and underground mining processes for several different commodities and support services are incorporated into the training. This training package includes two components:

- **Ergonomics and Risk Factor Awareness Training for Instructors** – Designed to give instructors sufficient information about ergonomics and risk factors to adequately present similar training to employees. It includes information that should allow the instructor to respond to questions regarding material included in training given to other employees. Anyone who has not had any training in ergonomics should take this training. It would also be helpful for individuals who have had training, but their training either did not apply specifically to mining or did not emphasize the reporting of risk factor exposures.

- **Guide to Conducting Ergonomics and Risk Factor Awareness Training** – This guide will assist the instructor in conducting the Ergonomics and Risk Factor Awareness Training. It includes slides, along with discussion notes, descriptions of demonstrations, equipment needed for demonstrations, and suggested references. Handout materials and training evaluation forms are also included in this guide.

Training Topics

Four modules covering topics relevant to gaining a basic understanding of ergonomics, as well as how it applies to mining jobs, are included in this training. A fifth module consists of interactive exercises that require trainees to apply what they learned in the previous modules.

- **Introduction to Ergonomics** – Defines ergonomics and provides two examples of using ergonomics to solve a problem. There is also a discussion on the benefits of using ergonomics to improve jobs.

- **Musculoskeletal Disorders** – Defines cumulative trauma disorders or work-related MSDs and describes how a worker may progress from experiencing discomfort to developing a permanent, debilitating injury. The discussion includes three examples of this type of disorder.

- **Risk Factors and Root Causes** – Presents descriptions and examples of ergonomic risk factors and their root causes. The four main risk factors receive special emphasis: forceful exertions, awkward postures, repetition, and vibration.

- **Prevention** – Includes information about ways to control risk factors using engineering controls, administrative controls, work practices, and PPE.

- **Exercises** – Give trainees an opportunity to apply the knowledge gained in the previous sections. The participants review a video and then identify risk factors, potential body parts affected, root causes, and potential methods for controlling the risk factors.

File Formats

This training is offered in two formats: electronic files and printed copies. The electronic files include:

- Ergonomics and Risk Factor Awareness Training for Instructors (Web and Adobe PDF formats)
- Materials needed for conducting Ergonomics and Risk Factor Awareness Training
    o Slide presentations (Microsoft PowerPoint and Adobe PDF formats)
    o Documents that may need to be copied such as handouts and evaluation forms (Microsoft Word, RTF, and Adobe PDF formats)
    o Discussion notes (Microsoft Word, RTF, and Adobe PDF formats)

The Ergonomics and Risk Factor Awareness Training for Miners (DHHS (NIOSH) Publication No. 2008–111, Information Circular (IC) 9497) may be ordered by contacting NIOSH at:

Telephone: 1–800–CDC–INFO (1–800–232–4636)
TTY: 1–888–232–6348
e-mail: cdcinfo@cdc.gov

or by downloading it from the NIOSH Mining Web site (www.cdc.gov/niosh/mining).

Ergonomics Observations:
Training for Behavior-based Safety Observers

When a behavior-based safety (BBS) system is part of an overall safety and health program, it may be helpful to also use this system to implement certain aspects of an ergonomics process, such as identifying and controlling some exposures to risk factors (those requiring simple changes, such as rearranging a work station or using a powered versus nonpowered handtool). If BBS observers are part of the ergonomics process, it will be necessary to define their role and to provide training appropriate for their role. For ergonomics observations, the observers will require training on identifying risk factor exposures. If the observers are expected to resolve some of the risk factor exposures, then they would require information on how to modify tasks, equipment, tools, work stations, environments, and methods using a hierarchical approach to controlling exposures (engineering controls, administrative controls, and PPE), with engineering controls being the preferred control measure [Chengalur et al. 2004]. The skill of resolving exposures may be assigned to just a few observers or other personnel responsible for implementing the ergonomics process. In this situation, resolving exposures may be done as a followup to an observation.

Objectives

The objective of ergonomics training for BBS observers is to improve their skills at identifying risk factor exposures, and then to give them sufficient information to resolve exposures if they are responsible for this as well. Because exposures are most often a result of poor design rather than methods or work practices (unsafe behaviors), the observers must be given basic knowledge on the design of tools, equipment, and work stations.

Content

The content of this training will depend on the role of the observers in the ergonomics process and on their role in the overall safety and health program as well. It should follow the observation process used by the observers to conduct safety observations. Because this training needs to be specific to the mine implementing the ergonomics process, the content should be based on tasks performed at this mine that would normally be observed by the BBS observers and should depict exposures as either individual or multiple risk factor exposures. A suggested outline for this training would include:

- Ergonomics and Risk Factor Awareness Training for Miners
- Review of risk factor exposures common to your site – In most cases, emphasis would be on forceful exertions, awkward postures, repetition, and vibration.
- Identification of risk factor exposures – This content should include several examples of workers performing a variety of tasks commonly occurring at your site. Include enough examples so the observers are comfortable with identifying exposures.
- Documentation of risk factor exposures – Observers should be given practice at documenting the exposures. If your site does not use a form that includes detailed information about the risk factors, then you may want to consider the Ergonomics Observations Form described in Section IV.

If the observers are also responsible for correcting the risk factor exposures, then the content should include basic design principles. Suggested training topics could include:

- Anthropometry
- Work station design
- Manual materials handling
- Ergonomics and seating
- Handtools

Format

This training needs to be interactive and simulate the actual observation process. If this training is done in a classroom, short videos of tasks can be used to practice identifying risk factor exposures. Role-playing exercises can also be conducted, with one trainee assuming the role of the employee shown in the video while other trainees serve as observers. If field exercises are possible, then trainees can conduct and document observations while employees actually perform tasks. Both the role playing and field exercises should use the actual observation process and documentation method followed when completing BBS observations.

References

ACGIH [2007a]. Threshold limit values and biological exposures indices. Cincinnati, OH: American Conference of Governmental Industrial Hygienists.

ACGIH [2007b]. Whole-body vibration. Documentation of the TLVs and BEIs with other worldwide occupational exposure values: 2007. Cincinnati, OH: American Conference of Governmental Industrial Hygienists.

Bernard BP, ed. [1997]. Musculoskeletal disorders and workplace factors: a critical review of epidemiologic evidence for work-related musculoskeletal disorders of the neck, upper extremity, and low back. Cincinnati, OH: U.S. Department of Health and Human Services, Centers for Disease Control and Prevention, National Institute for Occupational Safety and Health, DHHS (NIOSH) Publication No. 97–141.

Bernard TE [2007]. Analysis tools for ergonomists. [http://personal.health.usf.edu/tbernard/ergotools/index.html]. Date accessed: June 2008.

Bovenzi M, Hulshof CT [1999]. An updated review of epidemiologic studies on the relationship between exposure to whole-body vibration and low back pain (1986–1997). Int Arch Occup Environ Health *72*(6):351–365.

Burgess-Limerick R, Straker L, Pollock C, Dennis G, Leveritt S, Johnson S [2007]. Implementation of the Participative Ergonomics for Manual tasks (PErforM) programme at four Australian underground coal mines. Int J Ind Ergon *37*(2):145–155.

Chengalur S, Rodgers S, Bernard T [2004]. Kodak's ergonomic design for people at work. 2nd ed. Hoboken, NJ: John Wiley and Sons, Inc., pp. 490–491.

Cohen AL, Gjessing CC, Fine LJ, Bernard BP, McGlothlin JD [1997]. Elements of ergonomics programs: a primer based on workplace evaluations of musculoskeletal disorders. Cincinnati, OH: U.S. Department of Health and Human Services, Centers for Disease Control and Prevention, National Institute for Occupational Safety and Health, DHHS (NIOSH) Publication No. 97–117.

Dul J [2003]. The strategic value of ergonomics for companies. Rotterdam, Netherlands: Erasmus University, Department of Management of Technology and Innovation. [http://ssrn.com/abstract=633883]. Date accessed: June 2008.

Feuerstein M, Miller VL, Burrell LM, Berger R [1998]. Occupational upper extremity disorders in the federal work force: prevalence, health care expenditures, and patterns of work disability. J Occup Environ Med *40*(6):546–555.

GAO [1997]. Worker protection: private sector ergonomics programs yield positive results. Washington, DC: U.S. General Accounting Office, report GAO/HEHS-97-163.

Haims MC, Carayon P [1998]. Theory and practice for the implementation of 'in-house', continuous improvement participatory ergonomic programs. Appl Ergon *29*(6):461–472.

Haines HM, Wilson JR [1998]. Development of a framework for participatory ergonomics. Norwich, U.K.: Health and Safety Executive.

Hignett S, McAtamney L [2000]. Rapid entire body assessment (REBA). Appl Ergon *31*:201–205.

Hudson P [2003]. Applying the lessons of high risk industries to health care. Qual Saf Health Care 12(Suppl 1):i7–i12.

IEA (International Ergonomics Association) [2008]. What is ergonomics. [http://www.iea.cc/browse.php?contID=what_is_ergonomics]. Date accessed: June 2008.

Knapschaefer J [1999]. Ergonomics and behavior-based safety: how some automotive companies are making it work. Safety + Health Mag (National Safety Council) *160*(4):58.

Krause TR [2002]. Myths, misconceptions, and wrong-headed ideas about behavior-based safety: why conventional wisdom is usually wrong. Ojai, CA: Behavioral Science Technology, Inc.

Kuorinka I, Jonsson B, Kilbom A, Vinterberg H, Biering-Sørensen F, Andersson G, Jørgensen K [1987]. Standardised Nordic questionnaires for the analysis of musculoskeletal symptoms. Appl Ergon *18*(3):233–237.

Laing A, Frazer M, Cole D, Kerr M, Wells R, Norman R [2005]. Study of the effectiveness of a participatory ergonomics intervention in reducing worker pain severity through physical exposure pathways. Ergonomics *48*:150–170.

Lings S, Leboeuf-Yde C [2000]. Whole-body vibration and low back pain: a systematic, critical review of the epidemiological literature 1992–1999. Int Arch Occup Environ Health *73*(5):290–297.

McAtamney L, Corlett EN [1993]. RULA: a survey method for the investigation of work-related upper limb disorders. Appl Ergon *24*(2):91–99.

Moore J, Garg A [1995]. The strain index: a proposed method to analyze jobs for risk of distal upper extremity disorders. AIHA J *56*(5):443–458.

National Research Council [1999]. Work-related musculoskeletal disorders: report, workshop summary, and workshop papers. Washington, DC: National Academy Press.

NIOSH [1989]. Criteria for a recommended standard: occupational exposure to hand-arm vibration. Cincinnati, OH: U.S. Department of Health and Human Services, Centers for Disease Control, National Institute for Occupational Safety and Health, Division of Standards Development and Technology Transfer, DHHS (NIOSH) Publication No. 89–106.

NIOSH [1997a]. Mining. Presentation by Tim Martin, Southern Ohio Coal Company, at NIOSH-sponsored Ergonomics Conference, Chicago, IL, January 8–9, 1997. [http://www.cdc.gov/niosh/topics/ergonomics/ewconf97/ec4marti.html]. Date accessed: June 2008.

NIOSH [1997b]. Plenary session I. Presentation by Thomas Albin of 3M at NIOSH-sponsored Ergonomics Conference, Chicago, IL, January 8–9, 1997. [http://www.cdc.gov/niosh/topics/ergonomics/EWconf97/ec4albin.html]. Date accessed: June 2008.

NIOSH, Cal/OSHA [2004]. Easy ergonomics: a guide to selecting non-powered hand tools. Cincinnati, OH: U.S. Department of Health and Human Services, Centers for Disease Control and Prevention, National Institute for Occupational Safety and Health, DHHS (NIOSH) Publication No. 2004–164.

NIOSH, Cal/OSHA, CNA Insurance Companies, and Material Handling Industry of America [2007]. Ergonomic guidelines for manual material handling. Cincinnati, OH: U.S. Department of Health and Human Services, Centers for Disease Control and Prevention, National Institute for Occupational Safety and Health, DHHS (NIOSH) Publication No. 2007–131.

OSHA [1995]. OSHA draft proposed ergonomic protection standard. Appendix A: Getting started. Washington, DC: U.S. Department of Labor, Occupational Safety and Health Administration.

OSHA [2008]. Computer workstations: good working positions. [http://www.osha.gov/SLTC/etools/computerworkstations/positions.html]. Date accessed: June 2008.

Scharf T, Vaught C, Kidd P, Steiner LJ, Kowalski KM, Wiehagen WJ, Rethi LL, Cole HP [2001]. Toward a typology of dynamic and hazardous work environments. Hum Ecol Risk Assess *7*(7):1827–1841.

Shell International [2003]. Winning hearts and minds: the road map. Rijswijk, Netherlands: Shell International Exploration and Production B.V. [http://www.energyinst.org.uk/heartsandminds/docs/roadmap.pdf]. Date accessed: June 2008.

Steiner LJ, Cornelius KM, Turin FC [1999]. Predicting system interactions in the design process. Am J Ind Med *Sep*(Suppl 1):58–60.

Sulzer-Azaroff B, Austin J [2000]. Does BBS work? Behavior-based safety and injury reduction: a survey of the evidence. Prof Saf *45*(7):19–24.

Torma-Krajewski J, Steiner LJ, Lewis P, Gust P, Johnson K [2006]. Ergonomics and mining: charting a path to a safer workplace. Pittsburgh, PA: U.S. Department of Health and Human Services, Centers for Disease Control and Prevention, National Institute for Occupational Safety and Health, DHHS (NIOSH) Publication No. 2006–141, IC 9491.

U.S. Bureau of Labor Statistics [1998]. BLS issues: 1996 lost-work time injuries and illnesses survey. Am Coll Occup Environ Med Rep *98*:6–7.

Warren N, Morse TF [2008]. Neutral posture. Farmington, CT: University of Connecticut Health Center, ErgoCenter. [http://www.oehc.uchc.edu/ergo_neutralposture.asp]. Date accessed: June 2008.

Washington State Department of Labor and Industries [2008a]. Caution zone checklist. [http://www.lni.wa.gov/wisha/ergo/evaltools/CautionZones2.pdf]. Date accessed: June 2008.

Washington State Department of Labor and Industries [2008b]. Hazard zone jobs checklist. [http://www.lni.wa.gov/wisha/ergo/evaltools/HazardZoneChecklist.PDF]. Date accessed: June 2008.

Waters TR, Putz-Anderson V, Garg A [1994]. Applications manual for the revised NIOSH lifting equation. Cincinnati, OH: U.S. Department of Health and Human Services, Centers for Disease Control and Prevention, National Institute for Occupational Safety and Health, DHHS (NIOSH) Publication No. 94–110.

Webster BS, Snook SH [1994]. The cost of compensable upper extremity cumulative trauma disorders. J Occup Med *36*(7):713–717.

Westrum R [1991]. Cultures with requisite imagination. In: Wise J, Stager P, Hopkin J, eds. Verification and validation in complex man-machine systems. New York: Springer.

Westrum R, Adamski A [1999]. Organizational factors associated with safety and mission success in aviation environments. In: Garland D, Wise J, Hopkin V, eds. Handbook of aviation human factors. Mahwah, NJ: Lawrence Erlbaum.

Wright B [2002]. Dealing for dollars: convincing managers that good ergonomics is good economics. Presented at the Second Annual Department of Defense Ergonomics Conference (Chantilly, VA, April 29–30, 2002).

Additional Sources About Ergonomics Processes

Alexander D, Orr G [1999]. Development of ergonomics programs. In: Karwowski W, Marras W, eds. The occupational ergonomics handbook. Boca Raton, FL: CRC Press LLC, pp. 79–94.

Butler MP [2003]. Corporate ergonomics programme at Scottish & Newcastle. Appl Ergon *34*(1):35–38.

de Jong AM, Vink P [2002]. Participatory ergonomics applied in installation work. Appl Ergon 33(5):439–448.

Gjessing CC, Schoenborn TF, Cohen A, eds. [1994]. Participatory ergonomic interventions in meatpacking plants. Cincinnati, OH: U.S. Department of Health and Human Services, Centers for Disease Control and Prevention, National Institute for Occupational Safety and Health, DHHS (NIOSH) Publication No. 94-124.

Hägg GM [2003]. Corporate initiatives in ergonomics: an introduction. Appl Ergon 34(1):3–15.

Hignett S [2001]. Embedding ergonomics in hospital culture: top-down and bottom-up strategies. Appl Ergon 32(1):61–69.

Joseph BS [2003]. Corporate ergonomics programme at Ford Motor Company. Appl Ergon 34(1):23–28.

McGlothlin W, Kaudewitz H, Wilmoth S [1999]. Implementation strategies for a corporate ergonomics directive: the Eastman Chemical Company story. In: Alexander D, ed. Applied ergonomics case studies 1. Norcross, GA: Engineering and Management Press, pp. 15–80.

Moore JS, Garg A [1998]. The effectiveness of participatory ergonomics in the red meat packing industry: evaluation of a corporation. Int J Ind Ergon 21(1):47–58.

Moreau M [2003]. Corporate ergonomics programme at automobiles Peugeot-Sochaux. Appl Ergon 34(1):29–34.

Munck-Ulfsfält U, Falck A, Forsberg A, Dahlin C, Eriksson A [2003]. Corporate ergonomics programme at Volvo Car Corporation. Appl Ergon 34(1):17–22.

OSHA [1990]. Ergonomics program management guidelines for meatpacking plants. Washington, DC: U.S. Department of Labor, Occupational Safety and Health Administration, OSHA report No. 3123.

Perry L [1997]. Implementing company-wide ergonomics. Workplace Ergon May/Jun:18–21.

Ridyard D, Hathaway J [2000]. The three dimensions of an ergonomics program. Occup Hazards 62(2):41–44.

Robinson D [2006]. Success with ergonomics: Gold Kist, Inc., Atlanta, GA. [http://www.osha.gov/dcsp/success_stories/ergonomics/goldkist.html]. Date accessed: June 2008.

Smyth J [2003]. Corporate ergonomics programme at BCM Airdrie. Appl Ergon 34(1):39–43.

Steiner LJ, James P, Turin F [2004]. Partnering for successful ergonomics: a study of musculoskeletal disorders in mining. Min Eng 56(11):39–43.

Torma-Krajewski J, Steiner LJ, Lewis P, Gust P, Johnson K [2007]. Implementation of an ergonomics process at a U.S. surface coal mine. Int J Ind Ergon 37(2):157–167.

Unger RL, Turin FC, Wiehagen WJ, Steiner LJ, Cornelius KM, Torma-Krajewski J [2002]. Initiating an ergonomics process at a surface coal mine. In: Bockosh GR, Kohler JL, Langton JF, Novak T, McCarter MK, Biviano A, eds. Proceedings of the 33rd Annual Institute on Mining Health, Safety and Research. Blacksburg, VA: Virginia Polytechnic Institute and State University, Department of Mining and Minerals Engineering, pp. 39–47.

U.S. Bureau of Labor Statistics [1991]. Occupational injuries and illnesses in the United States by industry, 1989. U.S. Department of Labor, Bureau of Labor Statistics, Bulletin 2379.

Zalk DM [2001]. Grassroots ergonomics: initiating an ergonomics program utilizing participatory techniques. Ann Occup Hyg 45(4):283–289.

APPENDIX

ERGONOMICS PROCESSES

Beyond traditional safety and health programs…

Ergonomics is a science serving to bridge production and safety. Its focus is straightforward—designing for a better fit between workers and the methods, tools, equipment, and work stations used by workers. A better fit results in safer and healthier jobs. If you are ready to move ahead and address ergonomic issues at your mine, it is important to first consider how you will do this to have an effective process. Some critical elements necessary for successful implementation and integration are provided here as a guide for getting started. Remember, it is important to plan for success.

A CHAMPION FOR ERGONOMICS

The role of a "champion" is to promote and serve as an advocate and leader in applying ergonomic principles for process improvement. Implementing a new process requires leadership. A champion serves in that capacity and works to demonstrate the value of process improvement. It involves a great deal of time, particularly at the beginning during periods of planning and implementation. The organization should support the champion by investing time to this effort. It will significantly increase the odds of successful implementation.

TRAINING

Training is an essential element of an ergonomics process as it results in risk factor identification and problem-solving skills. Management's active support and involvement in the training demonstrates their commitment and support for the process. Management training should show how ergonomics can be applied to reduce musculoskeletal disorder (MSD) risks while fostering a safer and healthier workplace. It also serves to demonstrate the value of the ergonomics process from a financial perspective. Successful examples from other companies can be used to demonstrate this point.

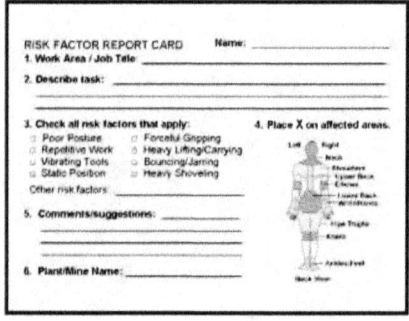

Just as with other types of safety hazards, employees should be skilled in recognizing significant risk factors and then be encouraged to report them. A simple card, as shown to the left, can be used by employees to identify and report exposures to MSD risk factors. Existing reporting forms can also be modified to include risk factors. Employees should be taught steps they can take individually to reduce their exposures, such as adjusting their work station to eliminate awkward postures. Training employees to report risk factor exposures is just the beginning.

IDENTIFYING / EVALUATING / CONTROLLING RISK FACTORS

To reduce exposures to risk factors, a procedure should be developed that ensures:

- Identification of risk factors
- Evaluation of risk factors in terms of root cause and level of exposure
- Development of solutions that reduce or eliminate risk factors

Once a risk factor exposure is identified, an ergonomics coordinator or team should then evaluate the risk factor exposure and determine the appropriate action for addressing the exposure. Remember, there can be many reasons for the exposure; consider the method, tools, equipment, work station, and environment. A procedure, as shown below, should be followed to ensure adequate evaluation and to determine an effective solution. Involving employees in the development of a solution will usually enhance the acceptance of the solution by the employees.

TRACKING PROGRESS

Quantifying the effectiveness of your ergonomics process depends strongly on the organization and the goal of the ergonomics process. It is common to see benefits measured in the number of work days lost, number of injuries/illnesses, number of near-misses, or changes in workers' compensation costs. But for some organizations, particularly small companies with limited injuries and illnesses, these measures may not be suitable. In such instances, use of survey tools, such as the Musculoskeletal Discomfort Survey form, may be more useful. Another constructive approach may be to quantify exposure levels of risk factors before and after implementing an intervention. For example, the distance an item is carried during a work shift may be measured before and after an intervention has been applied. Other examples include posture improvements, reducing the number of lifts completed, and reducing the number of repetitions performed. If you follow a behavior-based safety model, then risk factor exposures may be tracked with this system. As interventions are implemented, fewer exposures to risk factors should be seen.

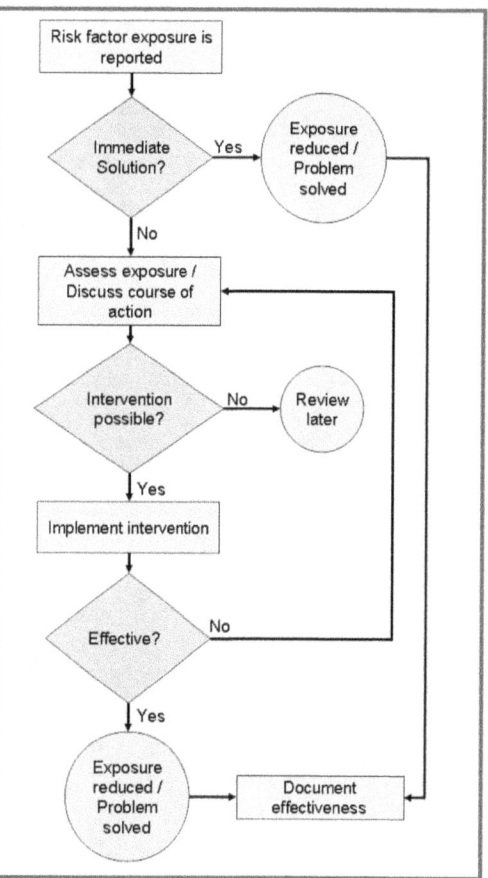

INTEGRATING ERGONOMICS

An ergonomics process can be implemented as a stand-alone activity or as an add-on to an existing process, such as a company's safety and health program. Regardless of the approach, it is important to fully maximize the effectiveness of the ergonomics process by integrating it with other processes that affect worker safety and health and the workplace. Examples of processes that could benefit from ergonomic input include:

- Purchasing new equipment and tools
- Purchasing personal protective equipment
- Designing new or modifying existing facilities, production lines, or work stations
- Determining work shifts and schedules
- Modifying work practices or procedures

Applying ergonomics to these processes at the planning stage will not only prevent the introduction of risk factors into the workplace, it will avoid costly reengineering efforts to correct situations. Incorporating ergonomics into the planning stage moves an ergonomics process from a *reactive* to a truly *proactive* mode.

www.ingramcontent.com/pod-product-compliance
Lightning Source LLC
Chambersburg PA
CBHW080254180526
45167CB00006B/2529